Canadian Special Publication of Fisheries and Aquatic Sciences 88

(personal copy ...) JEG

The Spiny Dogfish *(Squalus acanthias)* in the Northeast Pacific and a History of its Utilization

K. S. Ketchen

Department of Fisheries and Oceans
Fisheries Research Branch
Pacific Biological Station
Nanaimo, British Columbia V9R 5K6

DEPARTMENT OF FISHERIES AND OCEANS
Ottawa 1986

Published by	Publié par
Fisheries and Oceans	Pêches et Océans
Scientific Information and Publications Branch	Direction de l'information et des publications scientifiques

Ottawa K1A 0E6

Available from authorized bookstore agents, other bookstores
or you may send your prepaid order to the
Canadian Government Publishing Centre
Supply and Services Canada, Ottawa, Ont. K1A 0S9.

Make cheques or money orders payable in Canadian funds
to the Receiver General for Canada.

A deposit copy of this publication is also available
for reference in public libraries across Canada.

Canada: $8.00 — Catalogue No. Fs 41-31/88E
Other countries: $9.60 — ISBN 0-660-12049-6
ISSN 0706-6481

Price subject to change without notice

Director and Editor-in-Chief: J. Watson, Ph.D.
Editorial and Publishing Services: G. J. Neville
Typesetter: Nancy Poirier Typesetting Ltd., Ottawa, Ontario
Printer: Buchanan Printers, Winnipeg, Manitoba
Cover Design: Bayne-Herrera Graphic Communications Limited, Ottawa, Ontario
Cover Photograph: J. E. Ketcheson, Pacific Biological Station, Nanaimo, B.C.

This publication has been peer reviewed.

Correct citation for this publication:

KETCHEN, K. S. 1986. The spiny dogfish *(Squalus acanthias)* in the Northeast Pacific and a history of its utilization. Can. Spec. Publ. Fish. Aquat. Sci. 88: 78 p.

Contents

Abstract/Résumé iv

Chapter I. General Introduction 1
1. Purpose 1
2. Plan of Presentation 2

Chapter II. Background on the Spiny Dogfish 3
1. Common Names 3
2. Scientific Classification 3
3. Distribution in the Northeast Pacific 4
4. Migrations and Stock Delineation 7
5. Life-History Sketch 15

Chapter III. Prehistory of Spiny Dogfish Utilization 19
1. The Archaeological Record 19
2. First Contact and Beginning of the Historical Record 19

Chapter IV. History of the Commercial Fishery 25
1. Introduction 25
2. Lubrication and Lighting Era (1870–1916) 25
3. The Industrial Oil and Meal Era (1917–39) 32
4. The Great Vitamin Liver Fishery (1937–50) 35
5. The Government-Assisted Fishing Era (1951-74) 44
6. Coming of the Foodfish Era (1975 to present) 49

Chapter V. The Future 59
1. Industrial Prospects 59
2. The Foundation for a Management Policy 59

Acknowledgments 64

References 64

Appendices
1. Recaptures from American Tagging 68
2. Recaptures from Canadian Tagging 69
3. Calculation of Equivalent Round Weights 70
4. Marketed Quantities of Oil, 1876-1916 72
5. Landings of Dogfish and/or Oil, 1917–39 73
6. Problem of Assigning Landings to Area of Catch 74
7. Landed quantities of Dogfish and Liver 1940–54 75
8. Spiny Dogfish Removals from the Strait of Georgia, Jan.-Mar. 1957 75
9. Canadian Landings of Liver During 1954–70 76
10. Canadian Production by Subareas of Strait of Georgia, 1972–82 77
11. American Production by Subareas of Washington Internal Waters, 1976–80 78

Abstract

KETCHEN, K. S. 1986. The spiny dogfish *(Squalus acanthias)* in the Northeast Pacific and a history of its utilization. Can. Spec. Publ. Fish. Aquat. Sci. 88: 78 p.

This report reviews the taxonomy, distribution, migrations, definition of stocks, and life history of the spiny dogfish as background to an historical account of its utilization in the Northeast Pacific. Archaeological evidence traces the use of spiny dogfish by aboriginals from as early as 5000 B.P. to first contact with European explorers in the late 18th century. Its role in the culture of the famous Haida people of the Queen Charlotte Islands is summarized with illustrations.

The commercial history of utilization began before 1860 but remained undocumented until 1870. It passed through several phases: (i) the use of liver oil for lighting and lubrication (1870–1916) during which period production reached no less than 9000 t (equivalent round weight of dogfish); (ii) production of fish meal and oil for livestock and poultry industries (1917–39); (iii) production of vitamin A during the great liver fishery (1937–50) when yield of round dogfish exceeded 53 000 t; (iv) a period of uncertainty and government-subsidized operations (1951–74), and (v) the opening of foreign markets for spiny dogfish as food for human consumption (1975 to present).

In its latest phase, production exceeded 9000 t in 1979 but has fallen since that time partly because of reduced abundance in fishing areas closest to port, namely, those in Canadian and American internal waters of the Strait of Georgia and Puget Sound. Prospects for expansion of the fishery to currently unfished or lightly fished areas along the open coast of North America may depend on development of joint ventures with distant-water nations. If stability is a desired objective for existing and future fisheries, there is urgent need for development of an appropriate management policy by Canada and the United States.

Résumé

KETCHEN, K.S. 1986. The spiny dogfish *(Squalus acanthias)* in the Northeast Pacific and a history of its utilization. Can. Spec. Publ. Fish. Aquat. Sci. 88: 78 p.

Le présent rapport passe en revue la taxonomie, la distribution, les migrations, la définition des stocks et le cycle vital de l'aiguillat commun dans le contexte d'un compte rendu de son utilisation à travers les âges dans le nord-est du Pacifique. Des indications archéologiques font remonter l'utilisation de l'aiguillat commun par les autochtones à 5 000 ans avant notre ère jusqu'à leur premier contact avec les explorateurs européens à la fin du XVIIIe siècle. On résume à l'aide d'illustrations le rôle que ce poisson a joué dans la culture des célèbres indiens Haïdas des îles de la Reine-Charlotte.

On a commencé à pratiquer la pêche commerciale de l'aiguillat avant 1860, mais les prises n'ont pas été enregistrées avant 1870. Cette pêche a connu plusieurs phases : (i) l'utilisation de l'huile du foie pour l'éclairage et la lubrification (de 1870 à 1916), période au cours de laquelle la production a atteint au moins 9 000 t (poids entier équivalent d'aiguillat); (ii) la production de farine de poisson et d'huile pour l'élevage du bétail et de la volaille (de 1917 à 1939); (iii) la production de vitamine A lors de la grande pêche du foie (de 1937 à 1950) où les prises d'aiguillat entier ont dépassé 53 000 t; (iv) une période caractérisée par l'incertitude et des activités de pêche subventionnées par le gouvernement (de 1951 à 1974); et (v) l'ouverture de marchés étrangers en ce qui concerne l'aiguillat commun pour la consommation humaine (de 1975 à aujourd'hui).

Au cours de la dernière phase, la production a dépassé 9 000 t en 1979 mais a chuté depuis lors, en partie à cause de la baisse du nombre de poissons dans les pêcheries situées le plus près des ports, à savoir les secteurs situés dans les eaux intérieures canadiennes et américaines du détroit de Georgie et de la baie Puget. Les perspectives d'expansion de la pêche dans des secteurs actuellement inexploités ou peu pêchés le long de la côte ouverte de l'Amérique du Nord pourraient dépendre de la mise sur pied d'entreprises communes avec des pays où l'on pratique la pêche hauturière. Si l'on désire que les pêches actuelles et futures soient stables, il est urgent que le Canada et les États-Unis élaborent une politique de gestion adéquate.

Chapter I. General Introduction

The spiny dogfish *(Squalus acanthias)* shares with several species of Pacific salmon a recorded history in British Columbia fisheries statistics dating back to 1876. In that year the market value of spiny dogfish liver oil exported or used domestically was reportedly about equal to that of all canned and salted salmon (Anderson 1877). But there the comparison ends, for the long history of fishing for this species of small shark has been one of extreme volatility. At times the fishery has been nonexistent, at others worth millions of dollars but more commonly one of minor importance and incidental to those for other species. Whether in demand or not the spiny dogfish has been reviled by fishermen since early days as a despoiler of other, more valuable fishery resources or as a menace to fishing gear, and scorned by fishery officers, administrators, and collectors of statistics as nothing more than a trash fish — a curse upon mankind.

This disdain or, at best, indifference persisted even when, in 1944, the spiny dogfish became the fourth most important species in all of Canada's fisheries (after Atlantic cod, lobster, and herring) and first in British Columbia, in terms of landed value. In that year reference to this species by the western Chief Supervisor of Fisheries of the day warranted no more than two or three lines in an annual report of 28 pages devoted mostly to salmon. Today, as in the past, salmon fisheries understandably attract a great deal of attention, particularly those involving both commercial and sport fishermen, for the supply rarely if ever seems to exceed economic and recreational demand. Large numbers of fishermen and shoreworkers are involved in exploiting the salmon resource, to say nothing of the numbers of government employees charged with responsibility for monitoring and regulating the fisheries in the interest of conservation or pursuing the aims of enhancement.

In contrast, few Canadian fishermen today make even a part-time living from fishing for spiny dogfish and together with shoreworkers number less than 400 even under the best of economic conditions. For all but a few years in the long history of fishing for this species — during the 1940s when upwards of 3000 licences were issued — the numbers of fishermen and shoreworkers involved have always been far exceeded by those who were engaged in the salmon fisheries.

When resources deemed to be of high importance to the British Columbia fishery economy fall short of production expectations, biologists, fishery officers, fishery managers, and their political bosses are quick to feel the lash of public indignation. Again in sharp contrast, the only time there is hue and cry about the condition of the spiny dogfish resource (rarely, if ever dignified as such) is when its abundance appears to be too *high*. While the brief interval between the explosive development and subsequent collapse of the great fishery during the 1940s perhaps is an unfair example, alarms about overfishing were nonetheless not sounded by those whose livelihood was most affected. The attitude seemed to be, and still is, under somewhat less frantic circumstances: let's take all we can and let tomorrow take care of itself. In any event, for the majority of fishermen bent on making a living from other species, a noticeable reduction in abundance of spiny dogfish is cause for some rejoicing. Indeed it would not be a gross exaggeration to say that to speak seriously of the need for conservation of spiny dogfish is to incur suspicions of dementia.

1. Purpose

With such an unlikely candidate for a popularity contest, it would not be unreasonable of the reader to question the need for a history of the fishery for spiny dogfish. Actually there are several reasons whose collective importance covers a wide spectrum of interests that are not expected to appeal to all readers, but nevertheless deserve to be consolidated under one cover.

The first of these is to provide a convenient reference to an array of otherwise disorganized and out-of-date catch statistics forming the base from which fishery managers now face consideration of future management measures for spiny dogfish. As one of the consequences of Canada's move to extend her fisheries jurisdiction to 200 miles in 1977, management schemes have now been developed (in some cases hurriedly) for all of the purely marine resources in waters off British Columbia. For most species at least provisional restrictions in the form of total allowable catches (TACs) have been recommended to the Pacific Region management authority by the Fisheries Research Branch (e.g. see Stocker 1981). Clearly, the exercise of setting TACs implies that one has a reasonably sound estimate of the amount of fish on the grounds and of the equilibrium yield, namely, the quantity that can be taken without causing the stock to dwindle down to some level where it ceases to be economically attractive. Current recommendations in respect to spiny dogfish, however, are perilously dependent on the partially and imperfectly documented performance of fisheries of the distant past (Ketchen 1969). Thus a thorough historical review serves a utilitarian purpose in that it enables retracing of "footsteps" to the basis for setting TACs and provides a new perspective from which to view current attempts to manage this fishery and, of course, to consider the need for modification of TACs.

As already mentioned, fisheries for spiny dogfish historically had low priority with administrators and aroused demands from industry for some sort of government intervention only when the abundance (nuisance level) was too high. Thus it is not surprising that research priority on spiny dogfish for long intervals of time was equally low and funds available for biological investigation either minimal or non-existent. Fortunately this situation has improved slightly in recent years, but it is readily apparent that, as a consequence of neglect, the basic biology of the spiny dogfish and its role in the community of creatures that occupy the coastal sea are poorly understood. Particularly dismal is our understanding of the impact of the spiny dogfish on salmon, herring, and numerous demersal species (flounders, cod, crabs, etc.). Such an enquiry is difficult to pursue without an accurate appreciation of man's impact on spiny dogfish stock levels. Here again we find some practical value in assembling from the fast-fading but occasionally colorful past an historical review of the fishery.

Of less practical significance but of no less appealing interest is the opportunity to link that history, i.e., the written records of an enterprise as seen by the so-called civilized world, with whatever can be gleaned from prehistory — the archaeological record — and from reports on first contact of Europeans with the aboriginal peoples. The role of the spiny dogfish in the day-to-day life and culture of these early coastal inhabitants is an integral part of any history of utilization. As will become readily apparent, information marking the flow of events even through the recent written history is frequently not a great deal more enlightening that that derived from kitchen middens and carvings on ancient totem poles. Opportunity to establish the authenticity and meaning of what has been inscribed by pen, axe or chisel, has all but disappeared, and thus the aim of reconstructing an accurate record of activities associated with the use of spiny dogfish and developing an understanding of those activities is no easy task. The reader should understand the necessity for a great deal of interpretation, where otherwise personal recollections of contemporary observers would have been much more reassuring.

There is nothing unique about this problem, because even if one were to attempt to write a thorough history of the high profile Canadian west coast salmon fisheries and the economic and regulatory forces that shaped their development, the principal actors have long since left the stage and the same general difficulties would be encountered. With passage of time an array of catch statistics and regulations eventually loses its meaning unless someone of the day had the forethought to pen a concomitant narrative.

2. Plan of Presentation

As background to the historical review it is fitting to say something about the spiny dogfish itself: how it gets its common name and where it fits in the taxonomy or classification of modern sharks. A summary of its biology and life history will provide information pertinent to appreciation of the way in which the species becomes accessible to and responds to commercial utilization.

We shall look at the distribution of the species in the world's oceans and not only define its commercial distribution in the northeastern Pacific but also consider why spiny dogfish distribute themselves the way they do. This leads into a brief discussion of what is known about the number of stocks or independent populations of spiny dogfish, for without some understanding of results of tagging studies it would be difficult if not impossible to appreciate the impact of a localized fishery in one area on those in other areas of the coast. From this summary it becomes obvious that the fishery in the British Columbia region, or for that matter in any region along the North American coast, cannot be examined effectively in isolation.

Following an examination of prehistoric evidence of the use of spiny dogfish up to the time of first contact with European explorers, we then tackle the historical record, tracing the various phases of utilization through which the species has gone during the past century, and the various efforts made to induce exploitation.

Finally, in looking beyond the present day to prospects for maintenance or expansion of the fishery, consideration will be given to some management options and their consequences.

Chapter II. Background on the Spiny Dogfish

1. Common Names

As already apparent, we shall be using the name "spiny dogfish" and this undoubtedly raises a question in the mind of laymen. The "spiny" refers to the spines to be found at the leading edge of the two dorsal fins (Fig. 1). However, to commercial or sport fishermen along the west coast of Canada and the United States, a dogfish is a dogfish and if it needs adjectival adornment at all, the word or words would be out of place in polite literature. Until a little more than a decade ago the species was usually referred to in formal publications simply as dogfish or at best Pacific dogfish. For a long time, however, both in Canada and the United States it was euphemistically identified as "grayfish" by those in government and industry who hoped this would increase its appeal as an edible product.

In the absence of more than one kind of "dogfish" in the northeastern Pacific, except for two species occasionally and incorrectly given that name in earlier years in California, there is no necessity to be more explicit. However on the global scene, there is need for a more distinctive name because the term "dogfish" is used loosely in Europe as a catch-all for a number of species of small sharks (Budker 1971). Further, in western North Atlantic waters there are two species (the black dogfish and Cuban dogfish) that are recognized by the American Fisheries Society (AFS 1970) as belonging to the same family as the spiny dogfish. In addition there are two belonging to quite different families: the chain dogfish and the smooth dogfish. Thus the need for the term "spiny dogfish." Even this is not universally accepted for in the United Kingdom the name "spurdog" is in frequent usage in formal publications, and the species masquerades as "rock salmon" in the retail fish trade. It appears in some international statistical bulletins as the "picked dogfish."

Derivation of the word "dogfish" itself deserves some comment. The prefix "dog" dates from ancient days in Europe, often being used to denote some plant or animal worthless or unfit for human consumption. Perhaps this also serves to explain why the chum salmon *(Oncorhynchus keta)* has had the vernacular name "dog salmon." In early days this species was regarded to be of poor quality, was often discarded, and was the last of the five species of salmon to become fully acceptable to the British Columbia market.

2. Scientific Classification

To avoid chaos in the accurate description of animals and plants of the world, all are identified by a Latin scientific name and assigned to appropriate genera, families, orders, etc., in accordance with a strict set of international rules, not only to distinguish clearly a particular species but also to show its probable evolutionary relationship to other species. This systematizing process has been going on since the days of its founder, Linnaeus (1707–78). It is still in rather bewildering disarray, especially in regard to sharks, and subject to seemingly endless debate. Readers not wishing to become entangled in Latin names and the scientific squabbles about the classification of fishes may safely bypass this section and proceed directly to Section 3, below.

In the nomenclature of modern sharks and rays (Euselachii, according to Compagno's 1973 classification) there are four superorders, one of which is Squalomorphii which in turn contains three orders. One of these is the Order Squaliformes which contains those sharks distinguished principally by the presence of two dorsal fins; the presence or absence of a spine along the leading edge of one or both dorsal fins; presence of five gill-slits and the absence of an anal fin (for a more complete description see Compagno 1973).

FIG 1. The spiny dogfish, *Squalus acanthias*.

One of two families comprising Squaliformes is Family Squalidae containing about 19 genera and 76 species. Numbers vary depending on the authority recognized, largely because of disagreement about groups which may owe their existence simply to the current lack of sufficient specimens to distinguish genetic differences from natural variability in morphology. About eight of the genera and 10 species have been provisionally identified in the eastern Pacific, including the Hawaiian Islands (Compagno 1984). One of the genera is *Squalus*, to which the spiny dogfish belongs, while the others, also inhabitants of relatively cool waters, occur in moderately deep water of 500–2000 m (Kato et al. 1967) and some are luminescent.

There is only one representative of the genus *Squalus* along the west coast of North America. It was first described as *S. suckleyi* (Girard) 1854, a species believed to be separate from that occurring in European waters, *S. acanthias* Linnaeus 1766.

The species was first identified in British Columbia waters as *Acanthias suckleyi* by Lord (1866), but was renamed *S. acanthias* by T. H. Bean in 1881 (Clemens and Wilby 1961). However, the original name *S. suckleyi* appeared to be most favored through the remainder of the 19th century and the first half of the 20th.

Shmidt (1950), reporting on work he had completed by 1936, questioned the distinctiveness of *S. suckleyi*, stating that specimens taken in the southern part of the Sea of Okhotsk in the northwestern Pacific were very similar to those of *S. acanthias* from the Barents Sea, an extension of the northeastern Atlantic. He tentatively accorded the Okhotsk Sea form subspecific rank as *S. acanthias suckleyi* (Girard), noting the need for further study based on more extensive sampling.

Clemens and Wilby (1946) chose to adhere to the original name *S. suckleyi* in the first edition of their book *Fishes of the Pacific Coast of Canada*. Bigelow and Schroeder (1948) in their work on sharks of the western North Atlantic (probably conducted during the early 1940s like that of Clemens and Wilby), noted that the North Pacific form was usually recognized as a distinct species, but that the chief alternative character supposedly separating *suckleyi* from *acanthias* (position of the first dorsal spine in relation to the pectoral fin), was an unacceptable criterion. These authors could find no other morphological differences and thus concluded that ". . .North Pacific and North Atlantic populations of the *acanthias* group have not differentiated themselves specifically during the period since their ranges became discontinuous."

Subsequent authors (Roedel and Ripley 1950; Okada 1966; and Kato et al. 1967) accepted *S. acanthias* as the North Pacific form but Clemens and Wilby (1961), for reasons unknown, remained with *S. suckleyi*. Not until Hart (1973) produced the third version of Canada's Pacific fishes was uniformity of nomenclature achieved in respect to the eastern North Pacific.

All authors including Jones and Green (1976), the most recent students of Pacific dogfish taxonomy, appear to agree that *S. acanthias* is the only representative of the genus in the region extending from southeastern Bering Sea to Baja California. In the northwestern Pacific it occurs from the southern Sea of Okhotsk (Shmidt 1950) to the southeastern coast of Korea (Okada 1966), or even into the Gulf of Chihli in northern China (Compagno 1984).

As to the species' occurrence elsewhere in the Pacific, Bigelow and Schroeder (1948) noted some uncertainty but went so far as to say that species falling within their "*acanthias* group" (determined by the position of fins and spines), occur in the temperate and boreal zones of the southern hemisphere: Straits of Magellan, Australia and New Zealand, as well as in the South Atlantic. Kato et al. (1967) regard the southern Chilean form, sometimes referred to as *S. lebruni*, as *S. acanthias* since its difference from the North Atlantic and North Pacific form lies only in the number of vertebrae. Compagno (1984) concurs with the view that *S. acanthias* is broadly distributed in south temperate latitudes.

Regarding other species of the genus *Squalus* in the Pacific, there is still some confusion because of the problem of synonymy. The latest review (Compagno 1984) suggests there are seven species in all: two *(S. blainvillei* and *S. japonicus)* confined to the northwestern Pacific; two *(S. megalops* and *S. mitsukurii)* in both the northwestern and southwestern sectors with the latter being found also in the central Pacific (Hawaiian Islands) and off Chile; one species *(S. asper)* confined to Hawaiian waters, and two species *(S. melanurus* and *S. rancureli)* restricted to the southwestern Pacific (New Caledonia and New Hebrides, respectively). Further re-arrangement may be necessary since Compagno (1984) notes that the name *S. fernandinus* has been used recently to identify *S. acanthias, S. blainvillei*, and *S. japonicus*. In turn these three names have been used to describe *S. mitsukurii*.

Uncertainties and inconsistencies in classification will remain until taxonomists of the world can get together on a definitive study. With modern electrophoretic (electrochemical) techniques to test for genetic affinities, much of this confusion would quickly evaporate.

3. Distribution in the Northeast Pacific

In describing the distribution of a species it is useful to identify not only the limits of its natural occurrence, namely, wherever it is found even if only as isolated or stray individuals, but also where it thrives in sufficient concentrations to attract commercial exploitation. We shall deal with the former first.

In the cold waters of Bering Sea and in the western reaches of the Gulf of Alaska the spiny dogfish is uncommon if not rare. In extensive surveys conducted by the U.S. National Marine Fisheries Service and other agencies, occurrence of isolated individuals in bottom trawl surveys in Bering Sea has been noted no farther north or west than 68°35′N; 176°25′W and among the Aleutian Islands no farther west than 53°10′N; 167°05′W (personal communication, Gary B. Smith, Northwest and Alaska Fisheries Center, Seattle). However, Okada and Kobayashi (1968) noted a more westerly occurrence in the vicinity of Attu Island (53°00′N; 177°00′W).

Progressing eastward along the North American coast the spiny dogfish has been encountered only occasionally by bottom trawls in waters between Unimak Pass at the western extremity of the Alaska Peninsula and Portlock Bank (151°W) in the central Gulf of Alaska (Gary B. Smith, personal communication). Both NMFS surveys and those conducted by the International Pacific Halibut Commission (Anon. 1964) revealed greater incidence progressing eastward from Portlock but they were still of minor consequence in the eastern Gulf of Alaska. Alverson et al. (1964) noted that the ". . .importance of dogfish in the elasmobranch (cartilaginous fish) community. . .dropped rapidly in waters north and west of Cape Spencer" (58°10′N).

The species appears to be most common in waters off British Columbia and Washington and declines in significance from there southward through Oregon and California. The extreme southern limit of occurrence in the northern hemisphere is in the vicinity San Martin Island, southern Baja California (30°31′N; 116°10′W) as pinpointed by a recaptured tagged individual (Holland 1957).

The *commercial range* of the spiny dogfish in the northeastern Pacific is best measured by information on the distribution of fishing grounds and ports of landing during the period of heaviest exploitation (1941–49). Northern Hecate Strait (ca. 54°30′N) appears to be the approximate northern limit of large commercial concentrations. Some fishing activity, possibly incidental to the set-line fishery for halibut in water off and among islands of the archipelago of southeastern Alaska (ca. 55°N–58°N) was recorded during the abovementioned years, but at no time approached that in Hecate Strait.

The center of abundance appeared to lie in the British Columbia–Washington region (48°N–54°N) (Fig. 2) including inshore waters of the Strait of Georgia and Puget Sound. Ports in this region received by far the heaviest landings of dogfish livers during the 1940s. Within the region are three areas of particularly heavy concentration that over the past century have contributed

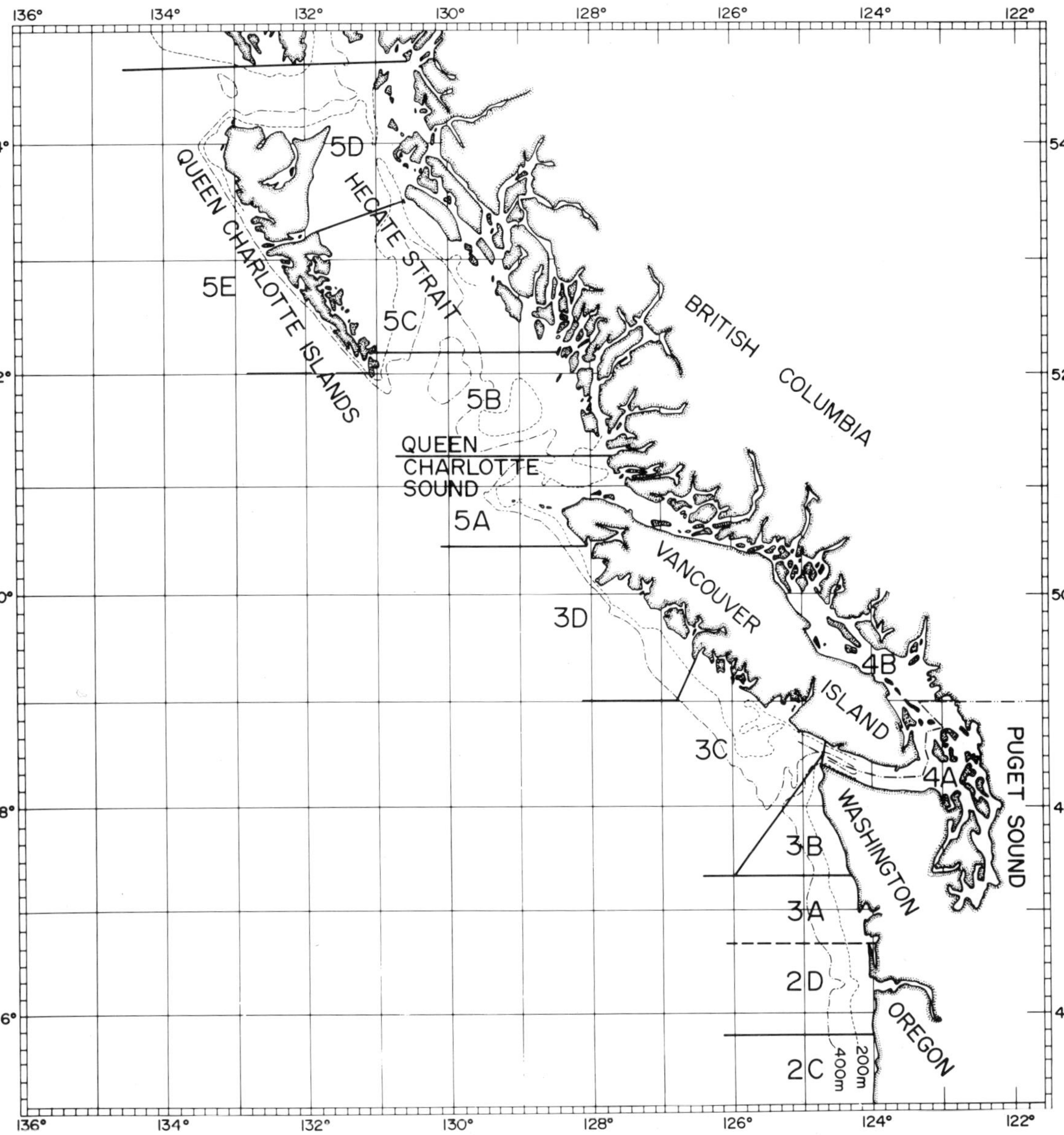

FIG 2. Coastal waters of British Columbia and Washington showing international (Canada–USA) groundfish statistical areas.

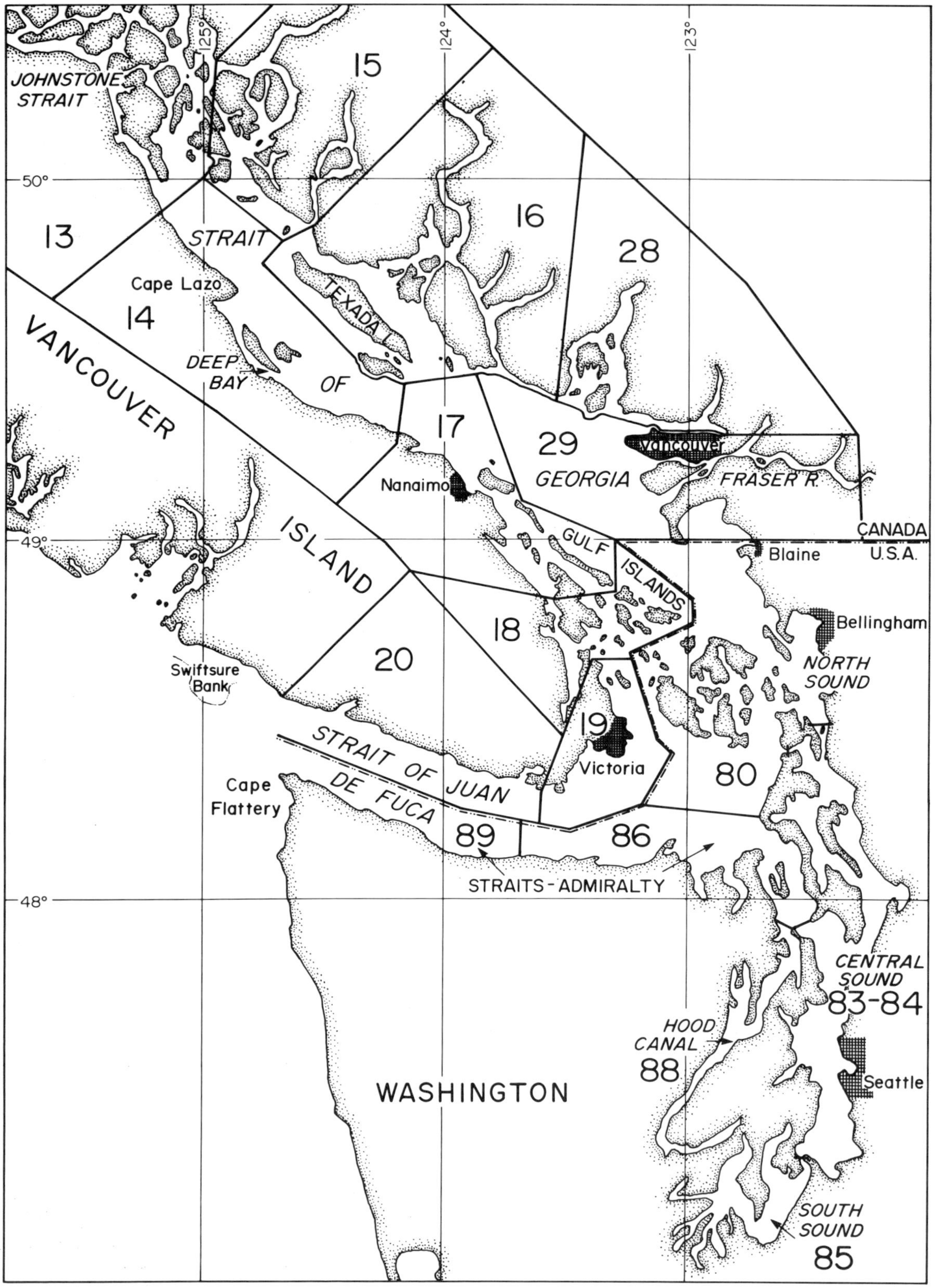

FIG 3. Minor statistical areas (MSA) in Canadian waters of the Straits of Georgia and Juan de Fuca (Area 4B) and in American waters of the same straits including Puget Sound (Area 4A).

most to commercial production in one form or another: (1) Hecate Strait, particularly the extensive "flats" that project eastward from shores of the Queen Charlotte Islands (known today as Groundfish Statistical Areas 5C and 5D), (2) the Strait of Georgia (Area 4B) lying between Vancouver Island and the mainland, and its southward confluence with Puget Sound[1] in the State of Washington (Area 4A), and (3) the lower west coast of Vancouver Island on and adjacent to La Pérouse and Swiftsure Banks (Area 3C) and south of Juan de Fuca Trench, off the coast of Washington, particularly between Destruction Island and Cape Flattery (Area 3B).

Historical records of landings in Oregon suggest the natural abundance of spiny dogfish was lower than off Washington but higher than off California. The southern limit of the commercial range, to judge from remarks of Roedel and Ripley (1950) was in the vicinity of Fort Bragg, California at 39°30′N Lat.

4. Migrations and Stock Delineation

Knowledge of the movements of spiny dogfish and their interrelationships along the west coast of North America is at best incomplete and probably will remain so until coordinated, coastwide tagging is undertaken. Various localized taggings have been conducted during the past four decades, some of considerable magnitude, particularly in the Strait of Georgia and Puget Sound. Holland (1957) reporting on results of releases in the latter area during the 1940s concluded that Puget Sound and both the American and Canadian portions of the Strait of Georgia supported indigenous populations. Similar conclusions could be drawn even from preliminary results of Canadian tagging in the Strait (Foerster 1942).

Between 1969 and 1972 over 17 000 dogfish were tagged in Puget Sound by the Washington State Department of Fisheries (WSDF) and students at the University of Washington. Of a total of 778 recaptures only four (0.5%) were recovered from waters off the open coast (Fujioka et al. 1974; M. Pederson, WSDF personal communication). Twenty three (3.0%) moved into Canadian waters of the Strait of Georgia (Fig. 4). These recapture rates probably underestimated the extent of emigration from Puget Sound because further returns (unpublished) were made after 1974, and except for 1973 there was negligible directed fishing for spiny dogfish either in Canadian inshore waters or offshore. This bias is perhaps illustrated by the fact that when American tagging was conducted on Swiftsure Bank at the entrance to Juan de Fuca Strait in June 1970, 60% of the recaptures were reported from grounds within the Strait of Georgia and Puget Sound. Still, the possibility of substantial immigration cannot be discounted.

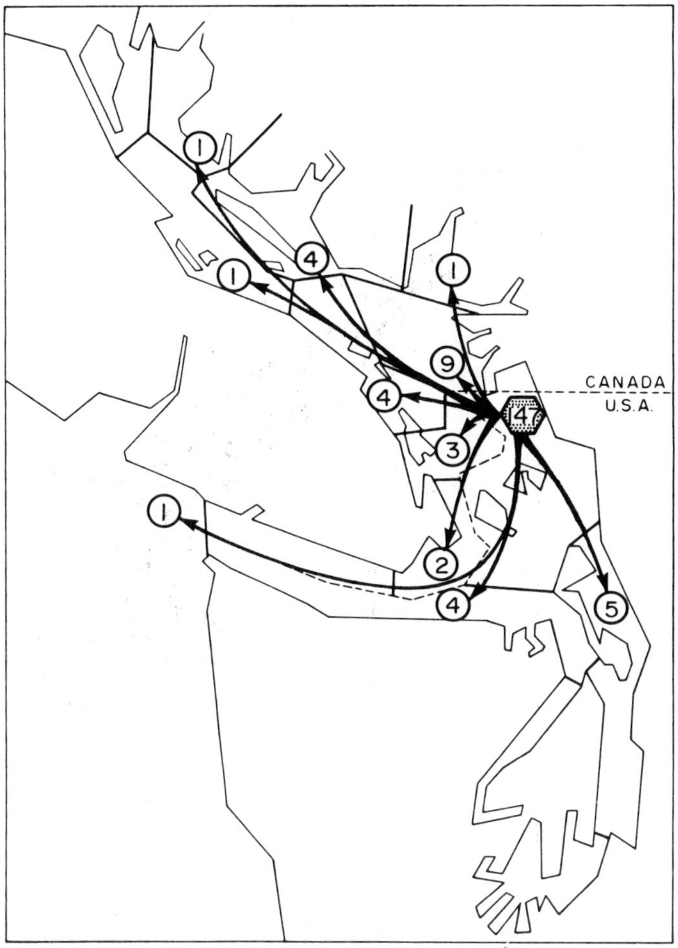

FIG 4. Movements of spiny dogfish as indicated by recaptures from tagging conducted in North Sound (MSA 80) or American waters of the Strait of Georgia (data from Appendix 1).

Preliminary results of Canadian tagging conducted in 1978, 1979, and 1980 have been reported by Brown et al. (1979), Beamish et al. (1981), and McFarlane et al. (1982), respectively, and are consolidated in Fig. 5. These results, including unpublished information from 1981 taggings (McFarlane, personal communication) are not unlike those obtained from the earlier Washington releases, in that only four (1.5%) of the tagged fish were recaptured off the open coast, despite considerably more directed commercial fishing there for spiny dogfish, i.e., more opportunity for recapture and discovery. A further similarity lies in the fact that considerable numbers (55 or 21%) were recaptured in adjacent American internal waters. Thus the more recent American and Canadian tagging results (with exception of those from the Swiftsure experiment) support Holland's (1957) view about the existence of populations in inshore waters that are largely independent of those off the open coast. It is necessary, however, to add a note of caution that most of the recaptures shown in Fig. 5 (and presumably in Fig. 4 also)

[1]Technically speaking the southern part of the Strait of Georgia extends into the United States internal waters, but to simplify later presentation, this segment of the Strait (subarea 80 in Fig. 3) will be identified with Puget Sound or called North Sound, one of the management or study areas currently used by the Washington State Department of Fisheries.

The Strait of Juan de Fuca forms part of both Areas 4A and 4B, and Johnstone Strait and Queen Charlotte Strait are included for general statistical purposes in Canadian Area 4B. However, none of these Straits has been or is a consistently major producer of spiny dogfish, although Alverson et al. (1964) noted high abundance during certain research cruises.

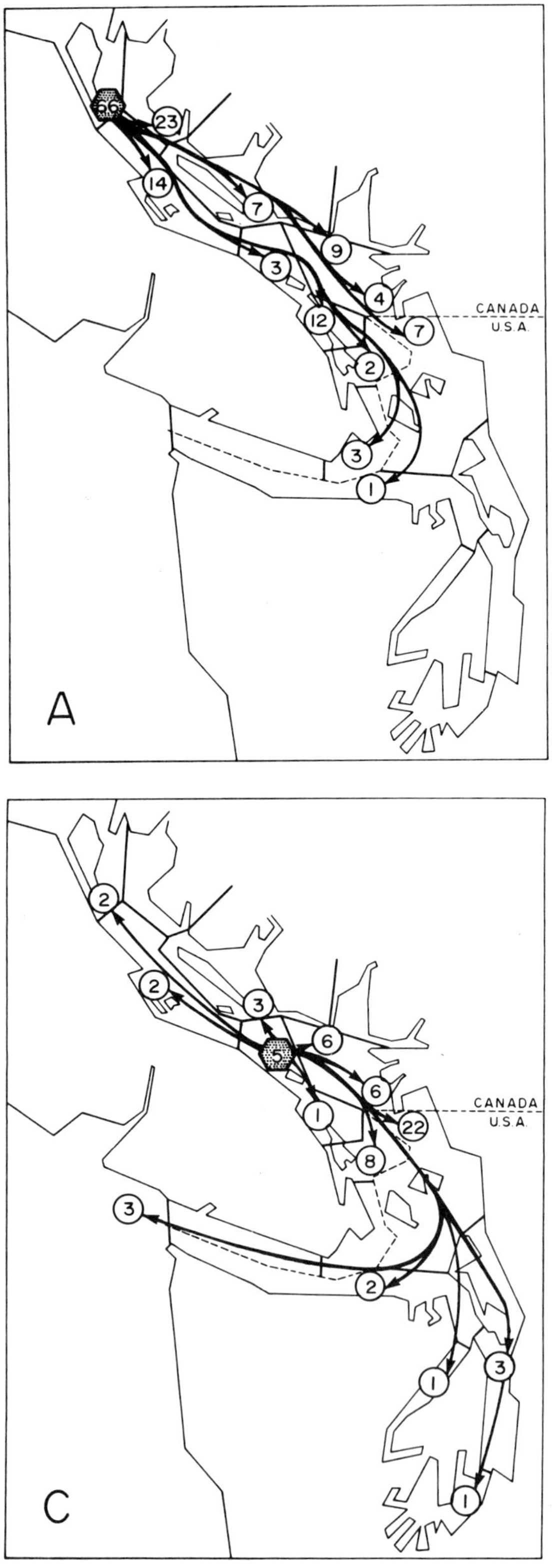

FIG 5. Movements of spiny dogfish as indicated by recaptures from tagging conducted in (A) Northern Strait (MSA 13), (B) Central Strait, mainland side (MSA 16), (C) Central Strait, Vancouver Island side (MSA 17A), and (D) Southern Strait, Gulf Islands (MSA 17B-18) (data from Appendix 2).

were made in the first 2 years after release. With the advent of more durable tags, it is likely longer-term analyses will reveal a greater dispersion not only within Canada–U.S. internal waters but also to waters off the open coast.

Knowledge of the number of spiny dogfish stocks and their patterns of migration along the open coast remains poor, despite the fact that during the 1940s nearly 2000 fish were tagged and released adjacent to Vancouver Island and Washington by WSDF (Bonham et al. 1949). An additional 900 were tagged by Canadian investigators in waters from Vancouver Island to Hecate Strait including some northern mainland inlets. Not until 1980–82 was Canadian tagging resumed, with a release of about 5700 individuals in Hecate Strait and Queen Charlotte Sound (McFarlane et al. 1982).

From the early and recent Canadian taggings, isolated recaptures have revealed spectacular movements, suggesting that certain stocks, parts of stocks, or at least some individual spiny dogfish are highly mobile. A large male dogfish tagged in northern Hecate Strait (54°04′N) was recovered 171 days later off Santa Cruz, California (37°10′N) (Manzer 1946), having covered a distance of 980 nautical miles (M) or 1810 km, extending from about the northern limit of the North American commercial range to the southern limit as defined here. Another long migration was reported by Holland (1957): from Ucluelet, Vancouver Island (48°55′N) to the aforementioned southern limit of the biological range off Baja California (30°30′N), or a distance of 1250 M (2320 km). Other recaptures from tagging off Vancouver Island were made at more moderate distances from Oregon to the latitude of San Francisco.

Much more remarkable were two trans-Pacific recaptures off Japan. Kauffman (1955) reported one was caught off the northern tip of Honshu 7 years after release in 1944 near Willapa Bay, Washington (Area 3A in Fig. 2). The other recapture was off the northeastern tip of Hokkaido after only 2 years at liberty from tagging in northern Hecate Strait (Areas 5D) in 1980 (McFarlane, personal communication). This evidence obviously helps to explain the existence of *Squalus acanthias* on both the western and eastern sides of the North Pacific, but poses a problem of interpretation vis-à-vis the evidence of sparse distribution of spiny dogfish from the northern extremity of British Columbia through the intervening Gulf of Alaska, Aleutian Islands, and Bering Sea. The minimum net distances travelled, had the two migrants followed the continental shelf, were about 4260 and 3610 M (7890 and 6690 km), respectively. Had they followed the Aleutian Chain to Kamchatka these distances could be reduced by 3 to 4%. On the other hand, even shorter distances would have been involved if they had taken a more southerly pelagic (near surface) route directly across the North Pacific, namely, about 3800 and 3370 M (7040 and 6240 km), respectively.

Support for the latter pathway is to be found in information on preferred temperatures (see below, p. 12) and the incidental catch records of Canadian, United States, and Japanese research vessels conducting surface gillnet and longline explorations for Pacific salmon on the high seas during summer months. Between 1954 and 1970 over 8700 sets were made within the broad area circumscribed by the broken line in Fig. 6 (data are from Powell and Pedersen (1957), Hanavan and Tanonaka (1959), and numerous unpublished documents submitted by Canada, Japan, and the United States to the International North Pacific Fisheries Commission). Encounters with spiny dogfish, most usually as solitary individuals, were made in 195 sets, positions of which are shown in the figure. An oceanic pathway along the general route of the Subarctic Current (Favorite et al. 1976) is suggested through latitudes where summer surface temperatures are in the neighborhood of 7°–12°C (Fig. 7). The relatively large number of encounters to the west of 180° should not be construed as a higher frequency of encounter, but rather as a reflection of the amount of fishing effort. Encounters west of 180° were 24 per 1000 sets compared with 18 east of 180°.

The point to be made is that the generally low frequency of occurrence rules out the likelihood that there is a significant ongoing east–west interchange.

Spectacular migrants aside, Holland (1957) observed from U.S. tagging results (along the open coast of North America) a tendency for dogfish tagged off the Washington and Vancouver Island coasts to migrate southward in fall and winter, and northward in spring and summer. Most winter recaptures were made 300–400 nautical miles (M) (480–650 km) south of the tagging sites while summer ones were mostly 0–100 M (0–160 km) north. So the annual latitudinal range of migration may be no more than 300–500 M (480–800 km).

The seasonal pattern of fishing as it occurred during the great liver fishery of the 1940s adds support to tagging evidence of a general north–south migration related to seasons of the year. A fishery for spiny dogfish occurred in waters off northern California in winter months only (Anon. 1949), suggesting a movement into that region from waters farther to the north. Off the Oregon coast, dogfish were fished when in greatest numbers during the fall, winter and spring (Cleaver 1951). A major "run" appeared to develop on grounds off the mouth of the Columbia River in the month of April (*Pacific Fisherman's News*, May 3, 1948), marking the usual beginning of the annual offshore fishery which then moved northward as the spring progressed. In Washington, landings of dogfish liver were greatest in the months of May through August (but included catches off British Columbia). Off the Vancouver Island coast mature dogfish (caught by sunken gill nets), were most abundant during the months of May through July while in Hecate Strait they appeared to be in highest abundance in the months of June through August (Barraclough 1948a). Thus, it is reasonable to conclude that some stocks or parts of stocks participate in a seasonal north–south movement.

Canadian results, though sparse, tend to confirm this conclusion. While some individuals may traverse the full commercial range of the species between summer and winter seasons, it is possible that these instances are more the exception than the rule. Certainly not all spiny dogfish engage in such lengthy seasonal movements, or even more moderate distances of 300–500 M, because it is known that, in the years 1945–49 when spiny dogfish were still a trawl-target species, substantial numbers occurred for example in Hecate Strait throughout the year, though

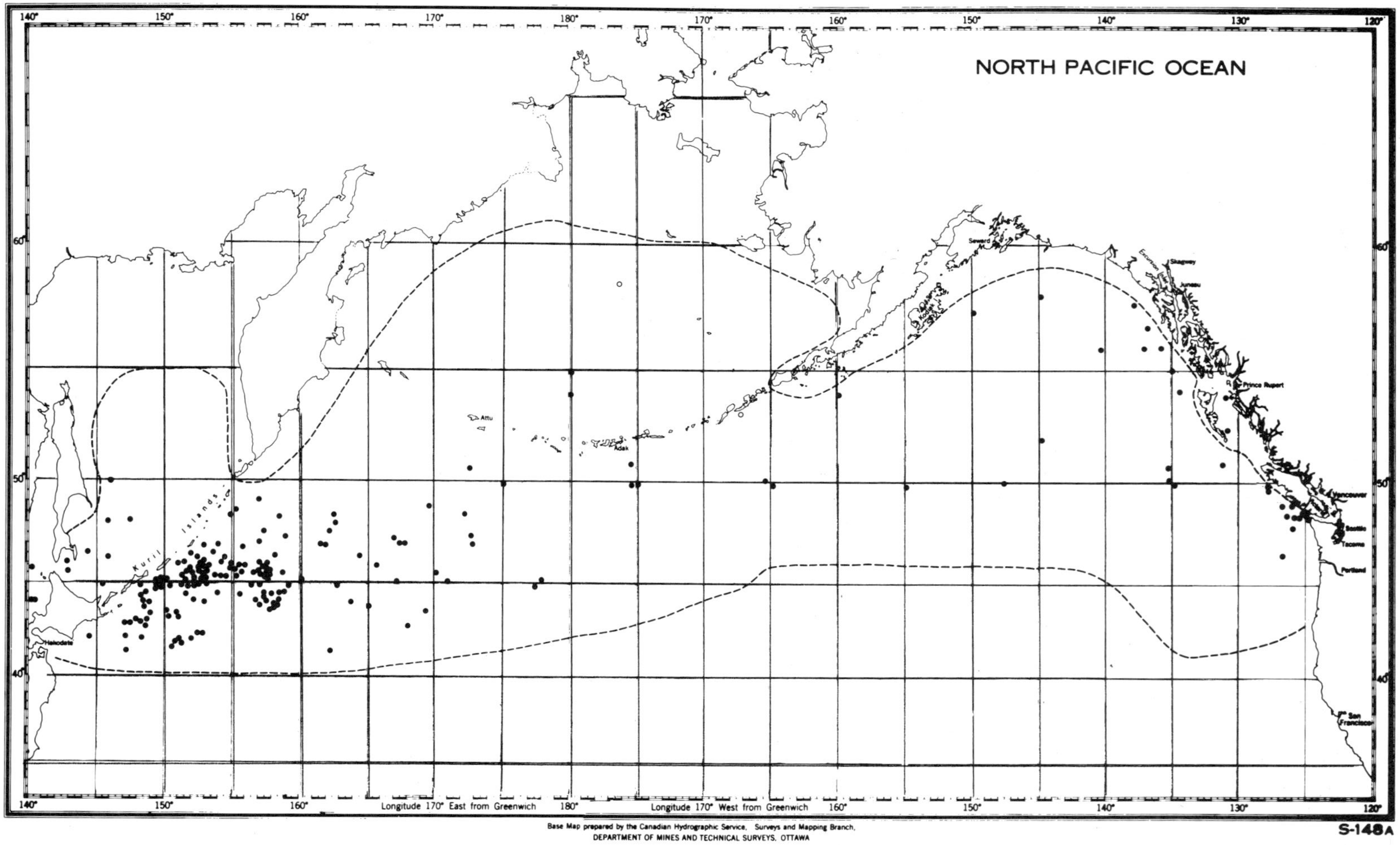

FIG 6. High-seas distribution of spiny dogfish in the North Pacific as indicated by incidental catches in surface gill net and longline surveys of salmon distribution within the region bounded by the broken line.

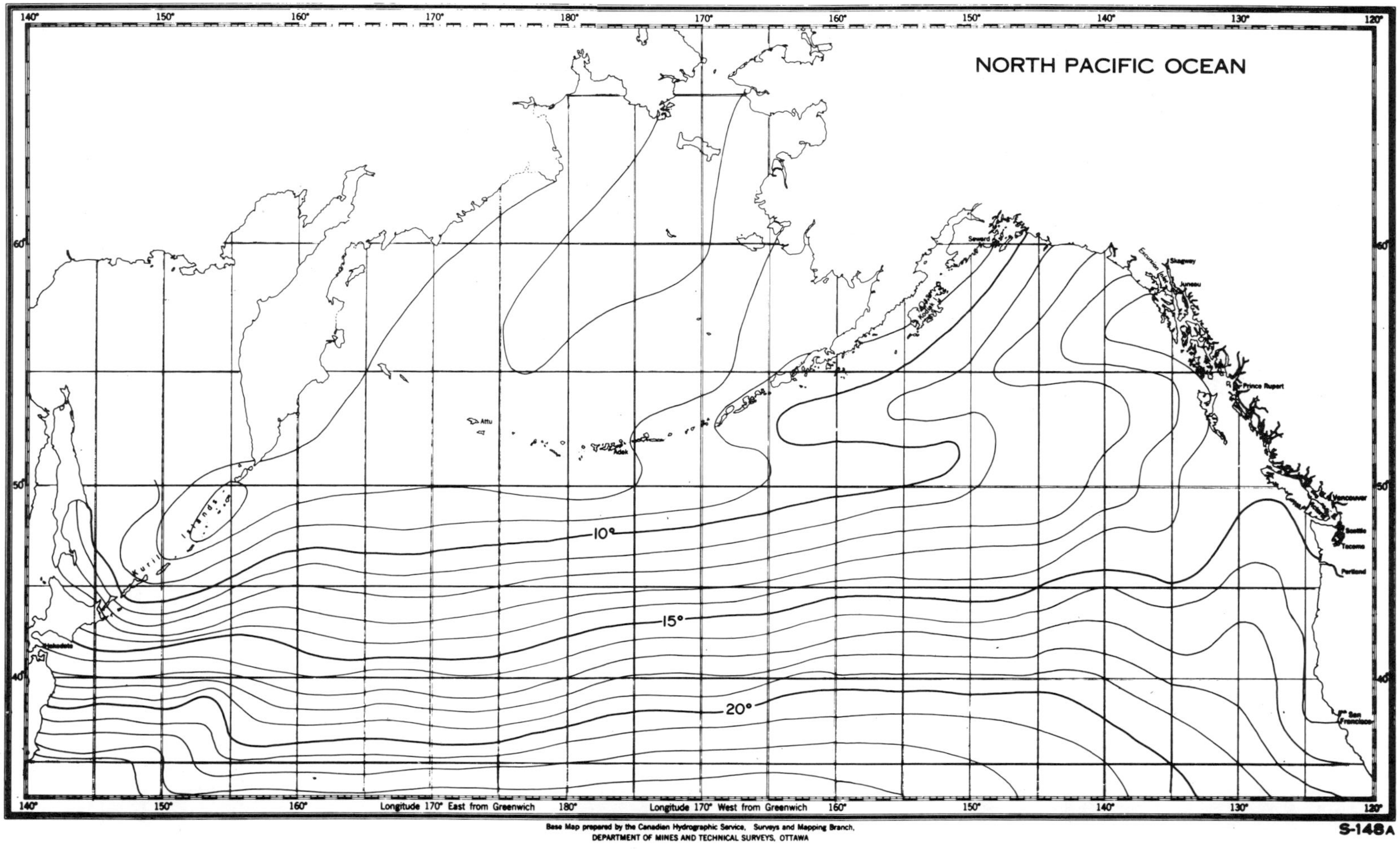

FIG 7. Long-term average sea-surface temperature (°C) of the North Pacific during the month of July (data from Gorshkov 1976).

on different and deeper grounds in winter (Fig. 8). The fact that production in winter months was but a small fraction of that in summer can be attributed mainly to bad weather conditions in winter, but it could also have been due in part to southward emigration from Hecate Strait.

Results of a small-scale tagging in November, 1944 may be indicative of the extent of the migratory pattern in Hecate Strait and Queen Charlotte Sound. Of 70 large spiny dogfish tagged in the inshore reaches of Queen Charlotte Sound (mainly in the entrance to Queen Charlotte Strait) four were recaptured in the 8th to 9th months after release (July–August) in the area of the major summer fishery on the Hecate Strait flats 150–200 M to the north (Fig. 9). Four other recaptures were made in the tagging area within 2 months of release, while two were made in the 3rd and 4th months, and one of these appeared to have been on its way to northern Hecate Strait. Thus, Queen Charlotte Strait may be the wintering area for many of the dogfish to be found on the Hecate Strait flats in summer.

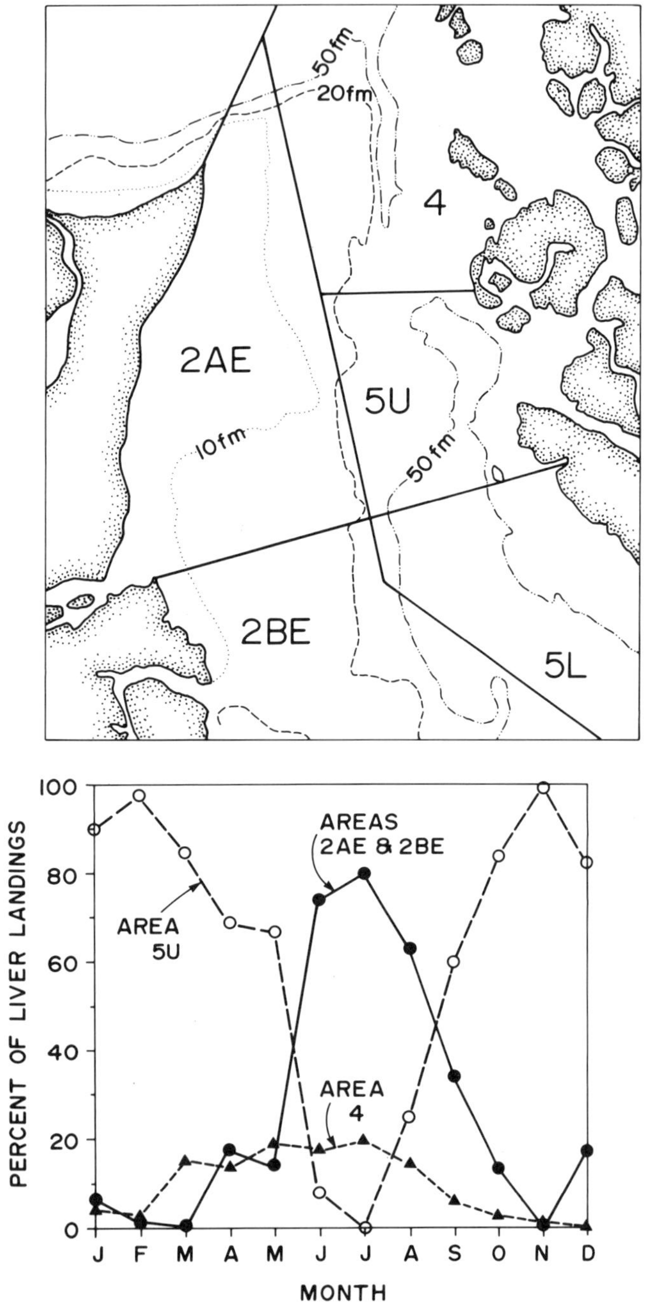

Fig 8. Minor statistical areas of Hecate Strait (above) and pattern of monthly average (1945–49) landings of spiny dogfish liver by Canadian trawlers (below) (data from Thomson and Yates 1961a and b).

Further support for this view is contained in evidence of the seasonal nature and location of the principal long-line (set-line) fishery during the 1946–49 period. Intensive fishing for spiny dogfish was confined almost exclusively to the November–April period and to waters of eastern Queen Charlotte Sound and the entrance to Queen Charlotte Strait (the general vicinity of the tagging site shown in Fig. 9).

Colloquial evidence of the concentrations of spiny dogfish in Queen Charlotte Strait prior to the great liver fishery of the 1940s is to be found in a remark of Syd Cooke, then editor of *Western Fisheries* magazine: "Clearing from Hardy Bay, and while on the approximate 39-mile course to Shushartie Bay, on the northern end of Vancouver Island, we ran through a dogfish school which extended for that distance, and as far as the eye could see to seaward." (*Western Fisheries*, Mar. 1946). However, relevance of this observation to the question of seasonal migrations is weakened by lack of reference to the time of year it was made.

Variations in the extent of seasonal migrations along the Pacific coast of the United States and Canada are in sharp contrast to those observed along the Atlantic coast. There the spiny dogfish is usually just a summer and early autumn visitor to the Gulf of Maine, Nova Scotia, the Gulf of St. Lawrence and Newfoundland (Leim and Scott 1966). While some may over-winter in deep water (Templeman 1963) most dogfish vacate these regions in late autumn in favor of more southerly waters between the latitudes of New York State and North Carolina.

A possible explanation for this more obvious seasonal shift in distribution is to be found in the seasonal patterns of change in water temperatures. According to Bigelow and Schroeder (1948), spiny dogfish occurring off the eastern United States appear to avoid waters colder than 6–7°C and warmer than 12–15°C. From the Gulf of Maine northward to waters off Canada temperatures on the shelf fall far below 6–7°C in winter.

If the same temperatures affect the presence or absence of spiny dogfish along the west coast of North America, it is immediately evident from Fig. 10 that within the commercial range from roughly 55°N to 39°N, temperatures of 7–10°C predominate at continental shelf depths in both summer and winter. Thus, pronounced coastwide migrations such as those of East Coast stocks to avoid cold water are "unnecessary" for Pacific coast stocks.

Figure 10 also serves to explain the evidence of sparse distribution to the north and west of British Columbia. Temperatures of 4–6°C prevail along the continental shelf of the Gulf of Alaska in winter months, while warmer water is confined mainly to depths no greater than 50 m in summer.

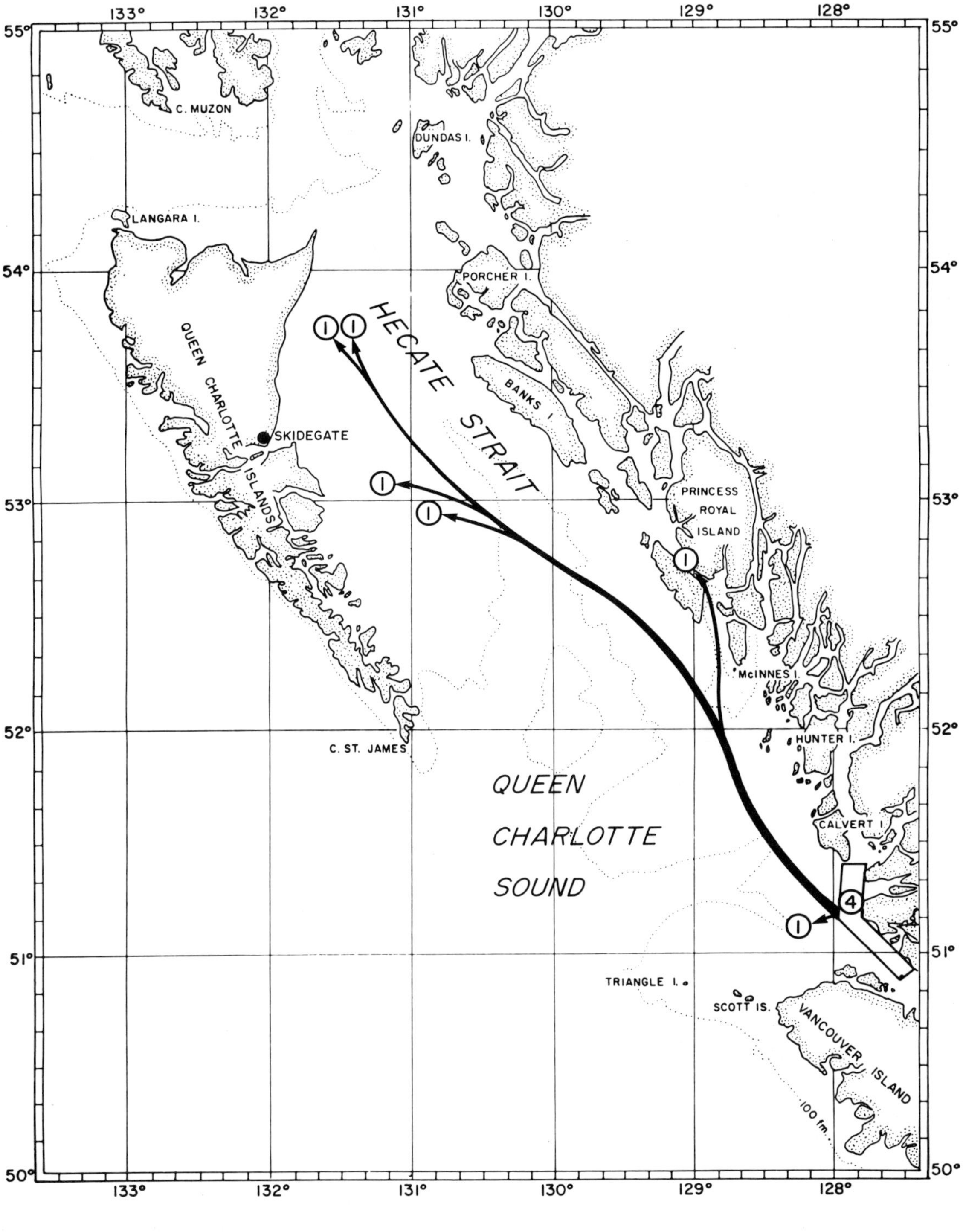

FIG 9. Results of tagging spiny dogfish at the entrance to Queen Charlotte Strait in November 1944. (Recaptures during first 9 months after release.)

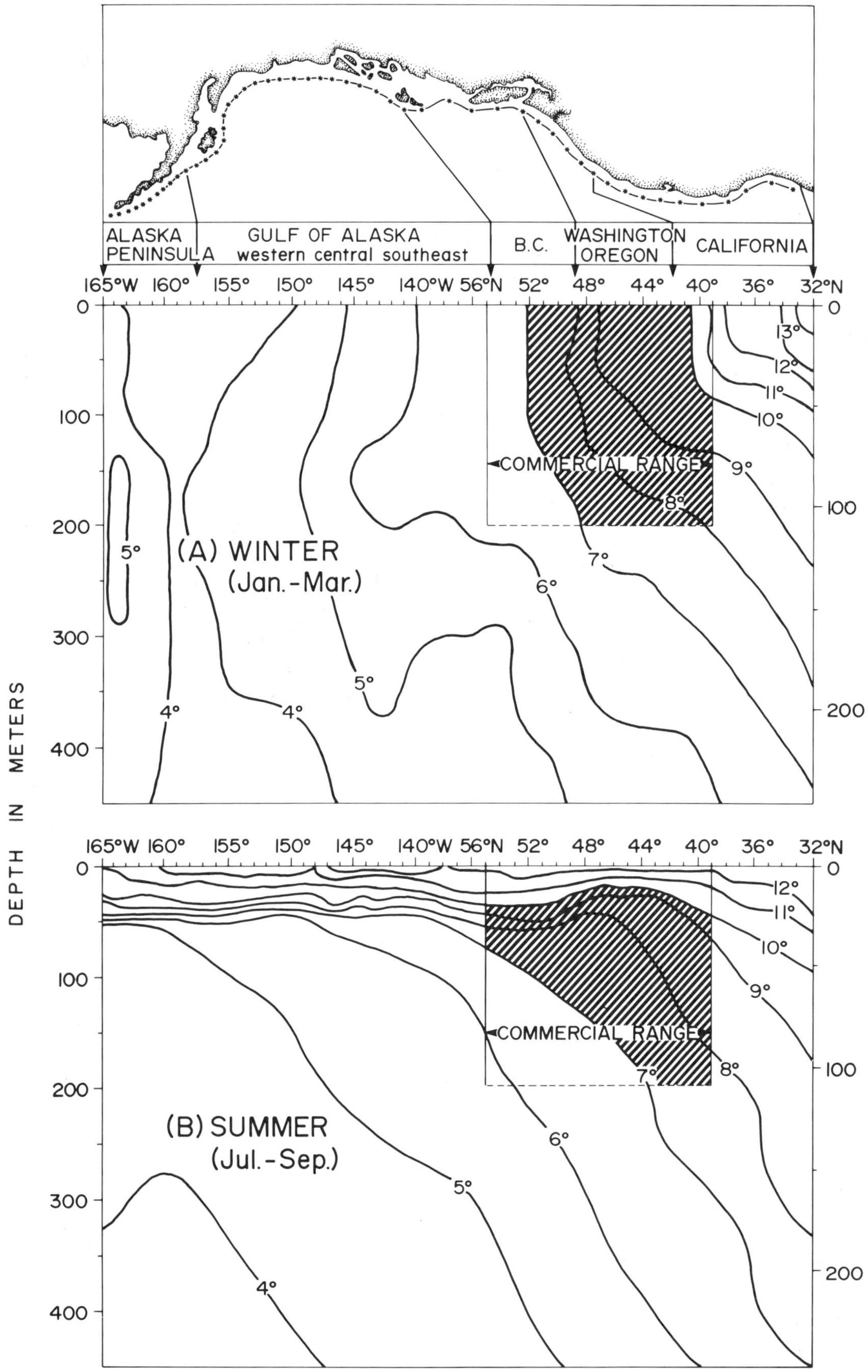

FIG 10. Summer and winter average temperature structure (°C) along the edge of the continental shelf in the Northeast Pacific Ocean. Data from Environment Canada, Marine Environmental Data Service, Data Rep. R8317176, covering the period 1956–65.

Roedel and Ripley (1950), commenting on the winter distribution of the spiny dogfish towards the southern limit of its range, noted that the species occurs at greater depth in southern than in northern California. This can be explained at least in part by the cline in winter water temperatures adjacent to that state at continental shelf depths (Fig. 10A).

To summarize in the context of this report, it may be said that while the historical fishery for dogfish in Canadian inshore waters (Strait of Georgia) appears to have involved stock(s) largely independent of those off the open coasts of Canada and USA, it cannot be considered in isolation from events in American internal waters, particularly the northern part of Puget Sound (North Sound) which is confluent with the Strait of Georgia. In offshore waters the situation is much less clear. Although fisheries off California probably had little impact on events in Hecate Strait (and vice versa), probably there was a gradient of impacts dependent on distance from Canadian waters.

5. Life-History Sketch

According to fossil records for the Americas, the physical characteristics of the genus *Squalus* of which the spiny dogfish is a member, have remained essentially unchanged since the Miocene or Oligocene epochs 30 to 40 million years ago. This means that the spiny dogfish has been successful in adapting to eons of environmental change, including changes in food supply and enemies. As a reflection of that success the species possesses a number of remarkable if not unique biological features.

a) Reproduction

Among the sharks and rays, fertilization takes place within the female's body, seminal fluid from the male being transmitted to the oviducts by means of a pair of "claspers" or appendages of the pelvic fins. The spiny dogfish of the Northeast Pacific breeds during the late fall and early winter, at which time large eggs 35 mm (nearly 1.5 in) in diameter and numbering two to 17 (Bonham et al. 1949) pass from the ovaries through the shell gland where they are simultaneously fertilized and encapsulated in a rubbery gelatinous "shell" before proceeding into the oviducts. There they remain for nearly 2 years (a gestation period unique to the animal kingdom). Ford (1921) was the first to note this remarkable phenomenon. After several months the shell dissolves leaving the embryos free, that is, unattached to the wall of the oviduct.

A schedule of reproductive stages (Table 1) shows that at any given time of year mature spiny dogfish will contain one or other of two stages of developing embryos. The casual observer and sometimes even long-experienced fishermen accordingly can be led to the mistaken belief that breeding and giving birth goes on year round.

Nourishment for the embryos is derived from the yolk material of the egg (Fig. 11). At the end of one year the average length of the embryos is 14–15 cm (ca. 5.75 in)

TABLE 1. Reproductive cycle in the female spiny dogfish (from Hart 1942).

Season	Fish no. 1		Fish no. 2	
		Year I		
Winter	Oviducts:	Minute embryos	Oviducts:	6-inch embryos
	Ovaries:	Small eggs	Ovaries:	Small eggs
Spring	Oviducts:	Small embryos	Oviducts:	7-inch embryos
	Ovaries:	Small eggs	Ovaries:	Eggs somewhat enlarged
Summer	Oviducts:	1-inch embryos	Oviducts:	8½-inch embryos
	Ovaries:	Small eggs	Ovaries:	Enlarging
Autumn	Oviducts:	3½-inch embryos	Oviducts:	9½-inch young almost ready for independent life
	Ovaries:	Small eggs	Ovaries:	Enlarged
		Year II		
Winter	Oviducts:	6-inch embryos	Oviducts:	Minute embryos
	Ovaries:	Small eggs	Ovaries:	Small eggs
Spring	Oviducts:	7-inch embryos	Oviducts:	Small embryos
	Ovaries:	eggs somewhat enlarged	Ovaries:	Small eggs
Summer	Oviducts:	8½-inch embryos	Oviducts:	1-inch embryos
	Ovaries:	Enlarging	Ovaries:	Small eggs
Autumn	Oviducts:	9½-inch young almost ready for independent life	Oviducts:	3½-inch embryos
	Ovaries:	Enlarged	Ovaries:	Small eggs

FIG 11. Spiny dogfish embryos about 18 cm in length (from Walsh 1984). In British Columbia waters they reach this size in mid-March, about 14 months after fertilization and 8 months prior to birth.

and a year later at full term, when the yolk has been completely absorbed, they average 26–27 cm (about 10.5 in) in length within a range of 23 to 30 cm. Numbers at birth, in British Columbia waters, vary from 2 to 16, depending on the size of the mother, but average between six and seven (Ketchen 1972). At birth the young are replicas of their parents and are ready to swim away, albeit rather weakly. The tips of the needle-sharp spines at this stage (and for the preceding year), are enclosed in skin thus protecting the yolk sacs of other embryos and protecting the mother while giving birth.

Release of the young appears to take place in midwater layers overlying depths of 165–350 m (90–200 fath). At least it is in this pelagic state that the youngest juveniles are found. Credit for this discovery should go to Capt. W. Kitzul Sr. of M/V *Sharlene K* who encountered new-born juveniles while midwater trawling at the entrance to Juan de Fuca Strait in March, 1973 at depths of 145–155 m overlying depths of 240–250 m. Beamish and Smith (1976) made the first scientific observation of this phenomenon, noting that during January in the Strait of Georgia new-born pups were to be found from 10 to 140 m off the bottom and 290–300 m from the surface (details from Beamish et al. 1978). Studies conducted in summer months revealed largest concentrations of young juveniles at depths 11–12 m from the surface (Beamish et al. 1982).

b) Growth and Survival

The spiny dogfish of the Northeast Pacific has an exceptionally slow growth rate. Age, as indicated by rings on the second dorsal spine, suggests that the female takes an average of 24 yr (range: 16–35) to reach maturity (Ketchen 1975). At that age she averages 94 cm (37 in) in total length and, if she survives, may take another 20 to 30 yr to reach maximum size of 130 cm (51 in). Males mature at an average age of 14 yr and achieve a maximum length of 107 cm (42 in) and age in excess of 40 yr (Ketchen 1972, 1975).

It is obvious from the above that, if one may be permitted to attribute (unscientifically) a human sense of purpose to animal behavior, the species is in no hurry to reproduce and assure its survival against predators and the elements. The reproductive rate of the female is in effect little more than three offspring per year and the process doesn't begin until she is more than 20 yr old. Prospects of survival are clearly good. Contrast this with the much less certain survival expectation of, say, the Pacific herring *(Clupea harengus pallasi)* which takes as few as three years to reach maturity and lays 20 000 to 40 000 eggs per year during its short life-span of 6–8 yr (Hourston and Haegele 1980). From the moment herring eggs are laid an intense struggle for survival begins as numerous predators along with adverse environmental factors take their toll. If only two of those 20 000 progeny survive to become spawning adults the stock will have been maintained for the next generation.

The natural death rate of the spiny dogfish appears to be less than 9% per year under conditions of natural equilibrium and possibly half that rate in exploited stocks (Wood et al. 1979). The causes of death are not clear. Cannibalism is one possibility for there are anecdotal accounts of spiny dogfish eating pups as they are being born and of older individuals attacking one another when cut or disembowled. However there is only one published scientific record (Bonham 1954) of a pup being found in the stomach of a spiny dogfish.

Predation is another possibility but it is suprising how few records we can find of species that will eat dogfish. There are unpublished records of the occasional presence of pups in the stomachs of lingcod *(Ophiodon elongatus)* and sablefish *(Anoplopoma fimbria)*, and an unconfirmed report of an adult being found in the stomach of a white shark *(Carcharodon carcharias)*.

The six-gill shark *(Hexachus griseus)* is known to eat adult dogfish that have become hooked on set-line gear (D. Miller, DFO, Pac. Biol. Stn. Nanaimo, personal communication), but whether they are consumed under natural conditions remains unknown. Spalding (1964) has reported them in the diet of northern sea lions *(Eumetopias jubata)*.

c) Food and Habitat

The spiny dogfish spends most of its early juvenile life in midwater where its food consists predominantly of a variety of small invertebrates but the diet changes to fish as the individual grows older and assumes more of a bottom-dwelling existence (Jones and Geen 1977). It is, however, an opportunitistic feeder, shifting from species to species as they are encountered. Bonham (1954) recorded at least 60 different food items in Washington waters with ratfish *(Hydrolagus colliei)* being most common.

However, any attempt to generalize on feeding habits is fraught with difficulty because so much depends on the locality, depth and time of year when sampling is conducted. In the Strait of Georgia, herring are prominent in the diet of dogfish during autumn months, especially along the former's migration route from waters off the open coast to inshore spawning areas. Yet at the same time of year dogfish may gather along Vancouver Island beaches where they are attracted by spawning capelin *(Mallotus villosus)*. In late winter some schools may be found at the mouth of the Fraser River feeding on post-spawning individuals of another member of the smelt family, the eulachon *(Thaleichthys pacificus)* (Chatwin and Forrester 1953).

While adult spiny dogfish are primarily fish-eaters, they frequently rise in the water column, even to the surface, to gorge themselves on swarms of "red feed" (euphausiids), the same small shrimp-like invertebrate that attracts baleen or filter-feeding whales.

Because of its relatively low metabolic rate the spiny dogfish digests its food slowly. Jones and Geen (1977) estimated from forced feeding with herring or salmon that the time between feedings is 16 days in British Columbia waters. Holden (1966), projecting from information on other species, estimated the time to be nine days in the North Sea. Fishermen interested in catching other species would, of course, dispute such conclusions, claiming that dogfish are feeding all the time! In any event, the impact of the spiny dogfish as a predator on other valuable species is a topic of such critical importance that further comment has been reserved for the concluding chapter.

Although adult and near-adult spiny dogfish do not confine themselves entirely to the ocean bottom, this is where commercial vessels generally catch them. Fixed gear such as baited set-lines or longlines and sunken gill nets tend to select for the larger sizes of fish. Otter-trawls or drag nets are generally much less selective.

Concentrations of marketable dogfish are to be found mainly on the continental shelf, and tend to be in shallow water in summer (in some areas females are often found at depths of less than 10 fath (18 m) and deeper water in winter 50 fath (94 m) or more). Maximum depths inhabited vary from area to area, occurring at over 200 fath (366 m) in the Strait of Georgia. Along the open coast Alverson et al. (1964) noted their presence (in trace quantities in the 200–299 fath (366–550 m) range off Oregon–Washington but none at that depth in the Gulf of Alaska, except in one of the coastal inlets (Prince William Sound).

d) Behavior

Although the occurrence of spiny dogfish often can be explained by the presence or absence of an attractive food supply, much of their behavior remains a mystery. Some writers describe the species as being nomadic or possessed of erratic as well as regular movements. Yet to a considerable extent these movements must be related in some way to the biennial breeding cycle of the female and to the presumably annual cycle of the male. Once the species reaches maturity, schools may be found consisting almost entirely of one sex or the other. Although the distance to which the two sexes may stray apart has never been determined, it is obvious that there is a well-timed convergence in the breeding season, for the occurrence of mature, non-pregnant females after that time, is a rarity.

Chapter III. Prehistory of Spiny Dogfish Utilization

Any account of the ways in which spiny dogfish have been utilized in the Northeast Pacific would be short-sighted if it were to exclude mention of prehistoric evidence. For this we turn first to the archaeologist for assistance in understanding what probably transpired during several thousands of years prior to the arrival of Europeans, and second to the ethnologist who endeavours to piece together customs observed following first contact, or recounted by village elders from personal memory and stories passed down from their ancestors.

1. The Archaeological Record

In Fladmark's (1982) introduction to the archaeology of British Columbia he deems it probable ". . . that most of the Northwest Coast has been ice-free and available for human occupation for about 12,000 years." While it is likely that habitation started about that time, the so-far-oldest dated, definite cultural assemblages for the region represent the period of about 9000–10 000 B.P. (before present).

Fladmark (1982) and others recognize a Lithic (stone) Stage characterized by the presence of "pebble-tool" and other stone artifact assemblages which lasted until about 5500 B.P. Because of the acidic condition of coastal soils preservation of organic materials such as bone is poor and in consequence there is little undisputed evidence of subsistence practices during the first 3500–4500 years of habitation.

There followed what is called the Developmental Stage which extended from 5500 to 5000 B.P. to "contact", i.e., to the arrival of early explorers and fur traders from Europe. This period is featured generally by the presence of large shell-middens, along the Pacific coast, from the Aleutian Islands to northern California, which date from around 4000–4500 B.P. (Fladmark 1982). Because of the buffering effect of shell deposits a large variety of organic implements and other remains have managed to survive the acidic condition of the coastal terrain, and among the earlier of these remains has been found evidence of fishing and marine mammal hunting technology.

Dorsal spines, the only clearly recognizable hard parts of a dogfish skeleton apart from teeth, appear to be of common occurrence in shell-midden sites on the British Columbia coast. For example, near the present mouth of the Fraser River, Dr. K. Fladmark (Simon Fraser Univ., personal communication) quoting Matson et al. (1976), states that spines were found at a level dated by radiocarbon analysis at roughly 2500–1500 B.P. On Mayne Island, one of the Gulf Islands shown in Area 18 in Fig. 3, spines were found at all levels, dating as far back as 4000 B.P.

On the west coast of Vancouver Island at Yuquot, better known as Friendly Cove where Captain James Cook landed in 1778, over 9500 spines were found in a small portion (4.4% by area) of the large shell-midden located there. Dewhirst (1980, and personal communication) noted that a small number (0.7%) of the specimens were found in a stratum dated 4600–4000 B.P.; 42% were in a 4000–2200 B.P. stratum; a smaller portion (21%) suggesting reduced usage of the site during 1200–300 B.P. and 36% during the historic period i.e., from first contact to 1966. Dewhirst suggests that the low number of spines in the first stratum (4600–4000 B.P.) was likely a result of poor conditions for preservation. These dates, placed in perhaps a more familiar historical time-frame, coincide with the period of the Egyptian Old Kingdom (Age of the Pyramids, 2700–2200 B.C.).

Towards the northern end of Vancouver Island near Port Hardy (adjacent to Queen Charlotte Strait), a few dogfish spines were found in a shell-free stratum dated possibly still older at 8020–4300 B.P. (Carlson 1979).

On the north coast of British Columbia, dogfish spines have been found in shell-middens in Prince Rupert Harbor, but remain undated. Regarding the Queen Charlotte Islands, Fladmark (1971) has exposed evidence of habitation since at least 8000 B.P., but advises (personal communication) that very little is yet known about the late prehistoric period (the last 4000 years). Many shell-middens are to be found and should have provided good preservation of dogfish spines and other bony material.

There is a dearth of ethnographic information to indicate what use if any was made of the spines themselves. Their natural sharpness certainly suggests they would have been useful as perforators or awls. Dewhirst (1980) classified 1% of his collection from Yuquot as having possibly been modified in shape for that purpose. It is known from historic and prehistoric times that the sandpaper-like skin of the spiny dogfish was used for polishing (see below), but there is also a possibility that the species was caught and used for other purposes, e.g., for the production of liver oil for various domestic needs, or for food during times of starvation, and that the spines were in most instances simply refuse from these enterprises. In any case, the use to which aboriginals put dogfish prior to the white man's arrival remains (in Dewhirst's words) "a thorny problem."

2. First Contact and Beginning of the Historical Record

a) Domestic use

Perhaps the earliest written reference to the spiny dogfish is an inferred one contained in the journal of Captain James Cook, who noted in 1778 that the Nootka used "fish skin" to polish wooden carvings and implements (Beaglehole 1967). If only because of its easy availability at Friendly Cove, the remark must refer to the spiny dogfish rather than to other, much less common species of shark. Captain Cook also made reference to face paints that appeared to be mixed in an oil, which again leads one to suspect spiny dogfish as the source.

Dawson (1880) quoting Captain George Dixon, one of the first explorers to contact the Haida of the Queen

Charlotte Islands, states in reference to a meeting off Skidegate (Skit-ei-get) village that

> "besides the large quantity of furs we got from this party . . .they had also a good quantity of oil in bladders of various sizes from a pint to a gallon, which we purchased for rings and buttons. This oil appeared to be of the most excellent kind for the lamp, was perfectly sweet, and chiefly collected from the fat of animals. . .".

Dawson remarks that ". . .the above mentioned was probably dog-fish liver oil, contained in the hollow bulb-shaped heads of the gigantic sea-tangle. . ." which he mistakenly identified as *Macrocystis* instead of *Nereocystis*. The use of the pneumatocyst or float of this kelp for storage of various kinds of fish oil continued to be in vogue long after the arrival of Europeans, for Lord (1866) reported that Indians of British Columbia's "north shore" (presumably the Tsimshian from the Nass and Skeena river areas) used this method for storage of eulachon oil. Three-foot lengths of the "great seawrack" could hold up to 3 pints of oil.

It is rather difficult to put much weight on Dawson's view that the oil offered in trade to Captain Dixon was obtained from dogfish. Indeed the term "sweet" is more suggestive of eulachon oil which may have been obtained in trade with people of the mainland coast.

There is a lack of direct information on the uses to which coastal Indians of various linguistic groupings put dogfish liver (and body) oil prior to the penetration of European customs into their way of life. Only recently has it been noted to have been used as a preservative for cedar canoes and as a waterproofing and softener of clothing woven from fine strands of cedar bark (Stewart 1984).

Possibly the oil was used in the cooking of other foods, or as an additive, but not likely as a food in itself. For the latter purpose, great stock was placed in eulachon oil, an important source of energy in diets of the Indian peoples along the Northwest Coast. In the region of British Columbia, eulachons were caught only in major mainland rivers in the territory of: the Tsimshian (Nass and Skeena rivers), the Bella Coola (Bella Coola River), possibly the Kwakiutl (other central coast rivers), and the Coast Salish (Fraser River). The Haida, Nootka, and the Carrier of the central interior obtained this oil through trade with the coastal mainland groups.

Dogfish liver oil may have been used in primitive oil-tanning of leather and in dressing or softening of sea

FIG 12. The Haida village fo Skidegate on the Queen Charlotte Islands (1878). Photograph courtesy of the Provincial Archives of British Columbia (Cat. No. 33786). Arrows point to several representations of the spiny dogfish on totem poles.

FIG 13. An entire totem pole *(center)* devoted to the spiny dogfish, and the base of a pole in the foreground *(left)* showing the traditional characterization of facial features. A Skidegate village photograph (ca. 1880) courtesy of the Ethnology Division, British Columbia Museum (Cat. No. PN 8055).

otter, seal, and sealion skins. Whether it ever played a role in illumination, in preference to eulachon oil remains a mystery.

Among the Straits Salish peoples of the southern Strait of Georgia and northern Puget Sound, before the arrival of the white man, Suttles (1951) determined that the Saanich people ate the flesh of the spiny dogfish in addition to making use of the oil and skin for other (unstated) purposes. However the custom may have been irregular and localized as it apparently did not extend to the adjoining Samish and Lummi peoples.

Once the sea otter resource of the Queen Charlotte Islands and other parts of the exposed British Columbia coast had been all but exterminated in the early 1800s, there was little recorded until the latter half of the century about Indian customs and culture, particularly in respect to their use of fish resources. Lord (1866), a naturalist and a rather uncharitable observer of the times, states:

> "... the dogfish is most useful and valuable to the Indians, who spear incredible numbers, split them and take out their livers. From these fatty livers a quantity of clear oil is extracted by heat and pressure applied in such a clumsy manner that at least one third is wasted." "I was credibly informed that one small tribe of Indians living on the west coast of Vancouver Island, by their bungling process of oil making, managed to obtain seven hundredweight of oil in one season: surely oil making alone would buy a company a handsome return for a judicious outlay of skill and capital. Several naval surgeons have assured me that they had fairly tested its curative powers — in diseases where oil is said to be efficaceous — and found it in every respect quite equal to the finest cod-liver oil."

Little did he realize how prophetic those remarks would become 8 to 9 decades later.

If Lord's information on this oil making activity and its location was correct, it was likely prompted by the demand of pioneers in the lumbering business for machine lubricants. The first sawmill on the west coast of Vancouver Island was established on the Alberni Canal in 1862 (Ireland 1954).

b) Role in Haida Culture

A more informative record of the role of the spiny dogfish in prehistoric life of coastal Indians is to be found in the well-documented cultural record of the Haida of the Queen Charlotte Islands. The dogfish or ka-hud-a as it was known in the Skidegate dialect (Dawson 1880) was

FIG 14. Two dimensional artistic impressions of the spiny dogfish reproduced with the kind permission of native artists: (1) Bill Reid, Vancouver, B.C., (2) Norman Tait, Vancouver, B.C., and (3) Robert Davidson, Whonnock, B.C. Reproductions from Stewart (1979).

FIG 15. A double mortuary pole with split-figure design of a spiny dogfish at the village of Skidegate ca. 1881. Photograph courtesy of the Ethnology Division, British Columbia Provincial Museum (Cat. No. PN 5795).

prominent as one of various crests represented by carvings on totem poles. The crest system was essentially the equivalent of the European heraldic system by which individuals or families indicated their rank and position in the social scale.

The use of animals as crests not only referred to the belief in man's descent from a common ancestor but was in accordance with Haida ". . .spirit theory that every animal was, or might be, the embodiment of a being who, at his own pleasure, could appear in the human form" (Swanton 1905). Thus animals of the ocean were called the ocean people, and within that category: the killer whale people, the herring people, the dogfish people, etc.

On the Queen Charlotte Islands and on part of the archipelago of southeastern Alaska the Haida were divided into two strictly exogamic clans — the people of the Raven and people of the Eagle. There were numerous families within the two clans and each family had the right to use a certain number of crests. The use of a particular crest was supposed to be the exclusive right of one clan but this was not strictly adhered to because of early day communication problems. Thus according to Swanton (1905) three of the 26 families within the Raven clan used the dogfish as a crest, but so also did two families in the Eagle clan.

In Swanton's (1905) classic contribution to the ethnology of the Haida, he illustrated by models two forms of the dogfish crest as represented on totem poles: one in which it occupies the top position on the pole, with tail projecting skyward and the head brought out forward; the other where the dogfish's head and face only are shown when ranked below other animals on the pole.

The stylized heterocercal tail of a dogfish forms the top of two poles seen in the earliest (1878) photograph of Skidegate village (Fig. 12). Another picture reveals a representation in which an entire pole is devoted to a dogfish complete with two dorsal fins and spines (Fig. 13), while details of the traditional dogfish face may be seen at the base of a pole in the foreground. Typically the features are distinguished by a triangular head with three (not five) gill-slits on each "cheek." These are depicted in the two-dimensional designs or tatoos of the dogfish as interpreted by several well-known modern native artists (Fig. 14). The split-figure design of Fig. 14(3) appears also on a double mortuary pole that stood at Skidegate more than 100 years ago (Fig. 15).

Chapter IV. History of the Commercial Fishery

1. Introduction

The commercial history of the fishery for dogfish (and here we are referring to the type of commerce that developed after the arrival of European explorers and settlers), is best examined by a number of stages whose irregularities disclose changing forms of utilization and changing format of reporting. First, 1870 to 1916 was a period characterized by extensive use of dogfish liver and body oil for various industrial lubrication requirements and for lighting purposes, and of the bodies themselves for production of fertilizer; second, 1917 to 1939, a period of use for oil and fishmeal in addition to or in place of fertilizer; third, 1937 to 1950, the period circumscribing the rapid development and equally rapid collapse of the great fishery for liver oil as a source of Vitamin A; fourth, 1951 to 1974, a lengthy interval of economic difficulty and of various attempts to resurrect a liver fishery or create a fishery for food with the aid of government subsidy programs; and finally, 1975 to present, the long-sought awakening of interest in the spiny dogfish as a source of food for human consumption and development of markets in Europe and the Orient.

The volatile economic history of dogfish usage in Canada is reflected in the frequent changes in format of landing statistics published over the years by the Canadian government (Table 2). This variability has presented a special challenge in reconstructing a cohesive panorama of the quantities of dogfish used during the more than 100-year history of the fishery in British Columbia and elsewhere along the North American coast. To achieve this it has been necessary to employ a number of conversion factors to express such products as liver, liver oil, body oil, etc., in terms of round weight. Previous authors in this regard have neither been consistent among one another nor within themselves, thus necessitating revision of many of the previous published records. Details of the method of calculating conversion factors and on problems of interpretation have been relegated to Appendix 3 to minimize the sometimes unavoidable soporific effect of dissertations of this kind. Suffice it to say that the historical record has never been without problems of interpretation due in part to the way in which dogfish products were and are marketed, and in part to the low priority placed on collection and documentation of landings. For many years the statistics were ambiguous, thus necessitating presentation of so-called minimum and maximum estimates of production in order to maintain continuity of the record and provide a perspective from which to view the heyday of the liver fishery in the 1940s and the foodfish fishery of the late 1970s and early 1980s. (The arithmetic gymnastics required to estimate minimum and maximum figures are described in Appendix 3.)

2. Lubrication and Lighting Era (1870–1916)

First reference to the spiny dogfish as an item of commercial enterprise did not appear in annual reports of the Dominion Government's Ministry of Marine and Fisheries until 1872. It was noted that in 1870 the steadily progressing fishery for dogfish (conducted entirely by Indians) exceeded in importance that of whaling, ". . .50,000 gallons of dogfish (liver) oil having been rendered, worth 40 cents a gallon" (Langevin 1872). In the same year an essayist, later to become Inspector of Fisheries for British Columbia, noted that among oil-producing fish, dogfish were ". . .so abundant as to give a lucrative employment to many fishermen and afford a boundless resource prospectively to others" (Anderson 1872).

In 1874 the Department's agent in British Columbia reported on the good quality of dogfish liver oil, noting that two lighthouses (Race Rocks near Victoria and the Fraser River lightship) were then burning dogfish liver oil exclusively ". . .giving a luminous and bright light, besides being cheaper (50 cents per gallon) than any other oil that can be imported" (Cooper 1874). In fact it was less than one-half the price of colza (rape seed) oil initially used at other lighthouses in the province. In spite of the potential savings, Cooper's recommendations went unheeded because plans had already been made to assure that lighthouses yet to be built (e.g., Cape Beale) would be fuelled even more expensively with oil derived from petroleum. It was evident, however, from Cooper's report for 1876 that there was at the time some uncertainty of supply respecting dogfish liver oil. In 1876, the failure to obtain a supply for the Race Rocks lighthouse from the area near Sooke was attributed to ". . .the migration of the Indians from that locality in consequence of the prevalence of smallpox among the tribe. . ." An alternative supplier for that year was found at Metlahkatlah Mission (near Prince Rupert). In any event dogfish liver oil continued to be used for the two lighthouses at least until 1884.

In the early 1870s it was evident that a large proportion of the liver oil being exported was destined for Great Britain and that strong hopes prevailed that the Washington Treaty of 1873 which restored benefits of the 1856 Reciprocity Treaty — among other things, the right of free access to U.S. markets — would be extended to British Columbia, the fledgling addition to the Dominion. Apparently this did not come to pass, for as late as 1883 (2 years before the United States abrogated the treaty), producers in British Columbia were still complaining about the "almost prohibitive" 25% ad valorem duty on imports of dogfish liver into the western United States and Washington territory.

Despite this impediment, the fishery for dogfish appeared to be developing rapidly by 1876.

> "The catching of these fish gives employment to a large number of persons along the sea-board of the Province; and the occupation will be a durable one, since the supply appears to be practically inexhaustible. Both to the native fisherman and the European, a valuable industry is thus opened and a large and wide circulation of cash is created" (Anderson 1877).

TABLE 2. Changes in reporting format of statistics on Canadian landing and marketing of spiny dogfish.

Period covered	Statistical format					
1875–1887	Dogfish oil refined	Dogfish seal & porpoise oil	Fish offal			
1888–1916	Fish oil	(Fish) fertilizer				
1917–1918	Dogfish caught and landed	Dogfish used fresh	Fish oil	Fish guano	Fish scrap or fertilizer	
1919–1929	Dogfish caught and landed	Fish oil	Fish meal	Fish fertilizer		
1930–1935	Dogfish					
	caught and landed	Marketed				
		Meal	Oil			
1936–1939	Dogfish					
	Caught and landed	Marketed				
		Livers	meal	Oil	Used fresh	
1940–1944	Dogfish					
	Caught and landed	Livers landed	Marketed			
			Livers	Meal	Liver oil	Body oil
1945–1946	Dogfish					
	Livers landed	Marketed				
		Liver oil	Livers			
1947–1954	Dogfish					
	Livers landed	Marketed				
		Liver oil				
	Dogfish					
1978–	Landed (rnd. wt)	Marketed				
		Fresh	Frozen	Flaps	Backs	Other

Production in 1876, the first year that fisheries statistics for British Columbia became available for routine inclusion in the annual reports of the Department of Marine and Fisheries, was reported as 275 000 litres (L) of liver oil. However this was simply the amount recorded in Custom House returns as having been exported to Great Britain at a value of $25,000 or 24% of the value of all fish exports (Whitcher 1877).

Anderson (1877) considered that in view of ". . .the large quantities consumed for lubricating and lighting purposes at the extensive saw-mills on Burrard Inlet and elsewhere, at the coal-mines at Nanaimo, Departure Bay, etc., and by the numerous steamers and sailing vessels frequenting these waters. . ." the total production of dogfish liver oil for the year (apparently almost entirely from the Strait of Georgia) was probably no less than three times the annual exported, namely, about 825 000 L. The equivalent round weight of dogfish represented by this amount of oil would depend on whether it was all liver oil or a mixture of liver and body oils. Following the procedures outlined in Appendix 3, we arrive at the minimum and maximum estimates of round weight for 1876 in Table 3 and Fig. 16, recognizing that they are at best merely informed guesses.

Anderson (1877), impressed by the potential of the dogfish fishery, went so far as to publish estimates of the cost of getting into the dogfish liver business. He considered that a two-man set-lining operation could be

TABLE 3. Estimated minimum and maximum equivalent round weights (tonnes) of spiny dogfish landed in British Columbia from fisheries adjacent to the Queen Charlotte Islands (Hecate Strait) and in the Strait of Georgia. Data derived from Appendix 4. (For details of calculation see Appendix 3 and text.)

	Queen Charlotte Islands		Strait of Georgia		British Columbia total			Queen Charlotte Islands		Strait of Georgia		British Columbia total	
Year	Min	Max	Min	Max	Min	Max	Year	Min	Max	Min	Max	Min	Max
1876	0		5775	9917	5775	9917	1897	1113	1911	511	877	1624	2788
77	0		3339	5734	3339	5734	98	1274	2188	735	1262	2009	3450
78	*		4802	8246	4802	8246	99	1392	2235	699	1292	2992	3438
79	709		3304	5673	4013	6382	1900	337	2296	700	1202	2037	3498
1880	1094		3787	6503	4881	7597	01	924	1587	1015	1743	1939	3330
81	1478		4515	7753	5993	9231	02	1113	1911	1239	2128	2352	4039
82	2019		6237	10710	8256	12729	03	1211	2079	2002	3438	3213	5517
83	2188		6909	11864	9097	14052	04	539	926	2198	3774	2737	4700
84	2464		315	541	2779	3005	1905	413	709	2352	4038	2765	4747
1885	2188		665	1142	2853	3330	06	511	877	2324	3991	2835	4868
86	1094		798	1370	1892	2464	07	252	433	2387	4099	2639	4532
87	2188		2135	3666	4323	4854	08	763	1310	2324	3991	3087	5301
88	637	1094	511	877	1148	1971	09	952	1635	2772	4760	3724	6395
89	952	1635	63	108	1015	1742	1910	1113	1911	413	709	1526	2620
1890	1176	2019	476	817	1652	2836	1911	952	1635	315	541	1267	2176
91	2135	3666	1148	1971	3283	5637	12	1113	1911	315	541	1428	2452
92	2226	3822	1435	2464	3661	6286	13	924	1587	*[a]	*	924	1587
93	2324	3991	1211	2079	3535	6070	14	0	0	*	*	0	0
94	2135	3666	1015	1743	3150	5409	1915	244	385	*	*	224	385
1895	1974	3390	1015	1743	2989	5133	16	511	877	248[b]	433	759	1310
96	987	1695	476	817	1463	2512							

[a]Asterisk indicates data missing from reports but landings known to be negligible.
[b]248 t landed for processing as food.

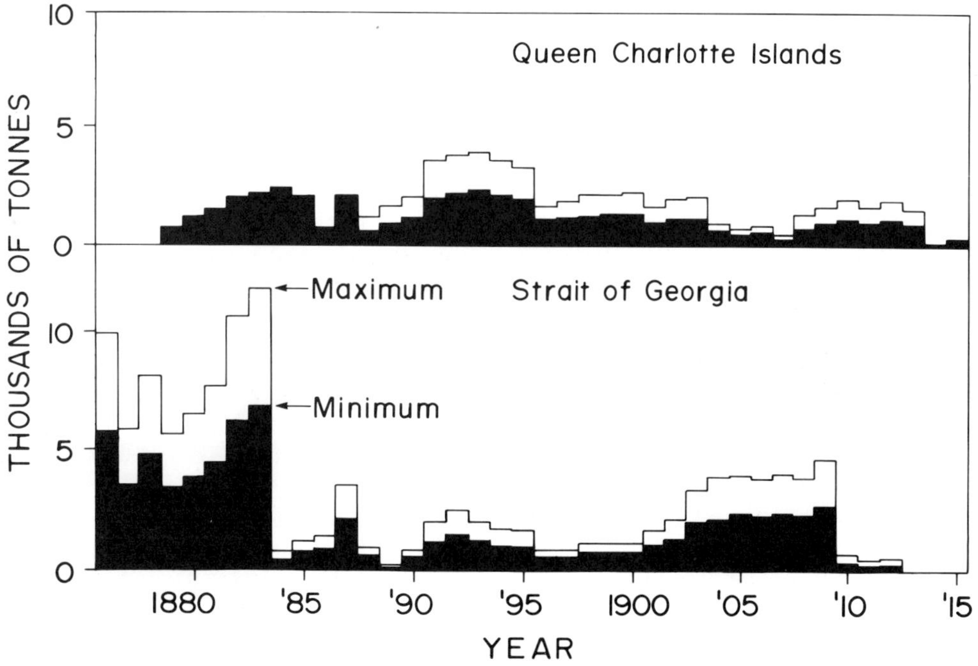

FIG 16. The 1875–1915 catch of spiny dogfish as estimated from commercial production records of liver oil from the Queen Charlotte Islands (Skidegate) and the Strait of Georgia (east coast of Vancouver Island). Data from Table 3.

launched with a capital outlay of only $250 (covering cost of boat, oars, sail, and net for catching bait-herring). He figured on an annual production per two-man team of 40 to 150 barrels (5000–20 000 L) with market value of $480 to $1800.

In 1877, production fell to less than 500 000 L, consisting of refined liver oil for export and "dogfish, seal and porpoise oil" (presumed to be almost entirely from dogfish and presumed to be Anderson's estimate of domestic consumption). To this point there was no clear reference in annual reports to production in the more remote northern areas of the province. However, Anderson (1878) noted that an attempt had been made to establish a processing plant at Skidegate on the Queen Charlotte Islands (see Fig. 9). This met with failure because ". . .white fishermen, at high wages, were employed, while the more economical services of the native fishermen were not utilized." Nevertheless it was evident that the Haida were producing oil on their own, for Dawson (1880) noted that it found ready sale among white traders and constituted one of the few remaining articles of legitimate value possessed by the natives. Elsewhere along the coast at many scattered points, other fishermen working for themselves with their own boats "prosecuted this industry with great success" (Anderson 1877). We have presumed that this referred primarily to the Strait of Georgia where mills, mines, and shipping were most active.

The aforementioned heavy customs duty imposed on imports of fishery products into the western USA, placed a damper on the legitimate trade in dogfish liver oil but at the same time provided considerable incentive for smuggling.

> "The Indians, tempted by the somewhat higher price which the traders of the opposite side can afford to pay, lose no opportunity of conveying their oil across the Strait to a dearer market. In this way, I am creditably informed, some 10,000 gallons or more were last year (1877) taken over to the vicinity of Neats Bay alone" (Anderson 1878).

Presumably he was referring to Neah Bay which lies immediately to the east of Cape Flattery at the entrance to Juan de Fuca Strait (Fig. 3).

Surely, no further evidence is needed to demonstrate that these early records of the dogfish fishery must be regarded as little more than rough approximations. Inspector Anderson had a staff of two: himself and a fishery overseer whose responsibilities were confined to the Fraser River salmon fishery. Thus it verges on the miraculous that he was in a position even to make rough approximations.

Alexander Caulfield Anderson (Fig. 17), in point of fact, was a remarkable individual. He was 62 years old when he became Inspector of Fisheries and had already distinguished himself in fur-trading, and exploring and mapping overland fur-trading routes through what would later become British Columbia. At the time of his retirement, after 27 years of service with the Hudson's Bay Company, he was a Chief Factor. From there he went on to become Collector of Customs for British Columbia, Postmaster of Victoria, Commissioner for settlement of Indian land claims, a notable historian and artist, before turning his attention to the world of fisheries. He died at the age of 70 apparently as a delayed result of overnight exposure on a Fraser River sandbar, while looking for a suitable site for a salmon hatchery.

FIG 17. Alexander C. Anderson, first Inspector of Fisheries for British Columbia (1876–84) and irrepressible promoter of a spiny dogfish fishing industry. Photograph courtesy of the Provincial Archives of British Columbia (Cat. No. 2228.)

In 1879, the Skidegate Oil Company, employing 16 native fishermen and five shoreworkers from Skidegate and the surrounding neighborhood in the Queen Charlotte Islands, was established with a capital investment of $25,000 for factory and boats. Although the operation did not get underway until summer was well advanced, the first year's production reached more than 270 000 L of liver oil (Anderson 1880). It was on this western side of Hecate Strait that Captain Dixon allegedly had been introduced to dogfish liver oil 100 years earlier and on the site that would become the base for a major liver fishery 65 years later.

The refining process used at Skidegate consisted of steaming the livers in a vat, with the oil extracted therefrom being reheated to separate the remaining water. The refined oil was then placed in 5-gallon cans for shipment.

For British Columbia as a whole, dogfish liver oil production reportedly reached 530 000 L, 34% of which was exported to London. So-called crude dogfish oil, probably consisting of body oil or a mixture of body and liver oils and produced by independent white and native fishermen, was in high demand at logging camps for lubrication of skid roads (Anderson 1881).

FIG 18. H.M.S. *Rocket*. Scene of the "Empire-shaking" lubrication experiment conducted in 1881 by Inspector of Fisheries A. C. Anderson. Photograph courtesy of the Provincial Archives of British Columbia (Cat. No. 7971).

By 1880, some of the 90 000 L of oil produced by the Skidegate Oil Co. was exported to Portland, Oregon where, as in San Francisco, it was well received. Exports to London rose to 250 000 L, while domestic consumption was estimated to be about 300 000 L.

Recalling the first attempt in 1874 to promote use of dogfish liver oil in all British Columbia lighthouses, another effort was made in 1881 with much grander visions of market potential but this time in respect to the virtues of the oil as a lubricant. Fisheries Inspector Anderson, while touring his domain aboard the naval vessel H.M.S. *Rocket* (Fig. 18), undertook with the cooperation of the chief engineer an experiment to compare the effectiveness of liver oil with the oil then in use by Her Majesty's navy:

> "The dogfish oil was applied to the starboard engines, whilst the port were worked with the ordinary service Rangoon oil (an imported vegetable product). The engines were driven at 140 revolutions, at which speed it was almost necessary to use a little water on the bearings with the Rangoon oil. This was found to be unnecessary on those lubricated with the dogfish oil. . . My opinion therefore is that as a lubricant the new oil is. . . equal if not superior to oil supplied to Her Majesty's ships. As the oil. . . could be supplied here at a cheaper cost, it may be hoped that the attention of the Admiralty will be drawn to it. Much encouragement would thus be given for the development of a local industry of practically unlimited extension" (Anderson 1881).

The one defect that the oil possessed, namely its disagreeable odor when warm, was subsequently diminished by the manufacturer (Skidegate Oil Co.) simply by further refinement (boiling).

Whether or not the Admiralty was ever presented with this recommendation is unknown. In any event nothing came of it, but dogfish continued to be the most important species contributing to production of fish oils, reaching 1 000 000 L in 1882 and an all-time high of 1 200 000 L in 1883. The latter figure was estimated to have been derived from 9 000 to 14 000 t of round dogfish depending on interpretation of the information provided (Table 3, Fig. 16).

The manager of the Skidegate Oil Co. reported in 1882 that his production had reached 40 000 gal. (182 000 L) of refined liver oil with ". . .no perceptible diminution in the quantity of fish (available)." (Anderson 1884). This production had been obtained from 400 000 dogfish landed in little over 4 months from waters along the east coast of the Queen Charlotte Islands. It is obvious from these statistics that the fishery was concentrating on exceptionally large fish, for some quick conversions of oil weight to liver weight, liver weight to round weight, to length, suggest that the average length of fish being caught was around 100 cm. This is not an unreasonable estimate, for three years after resumption of fishing (in 1946) on a stock that had remained unfished for almost three decades the average length was about 95 cm.

Although some rather limited markets for refined dogfish liver oil had been found in China and Hawaii, and although the oil received first prize at a Mechanics Fair in Portland, Oregon in 1883, the lack of free access to the U.S. market remained an obstacle to expansion of the Skidegate Oil Co. operation in Hecate Strait and production there leveled off in the 180 000–200 000 L range.

In an official report on exploration of the Queen Charlotte Islands prepared for the Government of British Columbia by N. W. Chittenden (quoted by Pittendrigh 1886), reference is made to the waters just outside the entrance to Skidegate Inlet as being ". . .the greatest known resort of the dogfish on the coast; the only place where they are caught continuously from spring until fall in large numbers". The true extent of dogfish concentrations on the Hecate Strait flats was not to be fully appreciated for another 60 years.

By 1888 production of refined oil at Skidegate had dropped to about half that achieved in 1884. This was due to temporarily poor economic conditions and not to localized depletion of the resource, for after a lapse of several years production reached new highs in the early 1890s (Fig. 16).

For reasons unknown the Skidegate Oil Co. was by then producing oil from both dogfish bodies and from livers, despite the fact that the former was of relatively poor quality. The Indians, in their domestic use of dogfish oil occasionally made use of the bodies:

> "After cleaning the fish, they cut them in pieces, boil them in vats, put them in large tubs, and the squaws press the oil out by tramping with their feet. This makes a very inferior oil which is mostly used for dressing of skins and on logging roads" (Mowat 1888).

Meanwhile, back in the Strait of Georgia, we encounter some difficulty with the interpretation of the landing statistics. Whereas production reached 987 000 L in 1883 after a steady rise from the 1870s, it allegedly fell abruptly in 1884 to only 45 000 L (or little more than 500 tons equivalent round weight (Fig. 16). This may have been an artifact resulting from a decline in the quality of the reporting system, for there was a change in fishery inspectors about that time (successors to Anderson probably lacked his zeal in promoting the cause of the dogfish industry). Alternatively and more charitably, they may have changed or abandoned attempts to estimate domestic production, which even under the best of circumstances must have been a hazardous exercise.

Still another problem emerged a few years later. Until 1887 statistics of the dogfish liver oil fishery made clear distinction between the refined oil produced at Skidegate and the combined production from all other areas. Starting in 1888 publication of production by locality was adopted, but unfortunately at the same time specific reference to dogfish oil was dropped and such records became lost in a catch-all category: "Fish Oils." Still, it remained possible to discern that the major dogfish oil producing areas (taking into account the newly developed salmon oil and offal plants that had opened on the Fraser River) were on the Queen Charlotte Islands side of Hecate Strait and along the east coast of Vancouver Island. This fits with present day knowledge of the areas of major production and thus enables us to follow general trends in activity of the dogfish fishery with some degree of confidence. This notwithstanding, there remained the possibility that efforts to keep a reasonably accurate account of domestic production had faltered or been abandoned.

Mowat (1890) stated that by 1889 ". . .home consumption (of dogfish oil) had increased at such a rate that present factories cannot supply the demand. . .". Yet this belies the statistical record, unless he was referring solely to Queen Charlotte Islands production (Fig. 16).

In any event, we enter a "dark age" (1888 to 1916) where there is scarce mention of dogfish in inspectors', and other reports. In 1892 the British Columbia Fisheries Commission made numerous recommendations pertaining to the discard of fish offal and carcasses, noting "that the system now prevailing along the coast of killing vast numbers of dogfish expressly for the use of the livers of said fish for oil purposes only, should be discontinued, unless the bodies of these fish are utilized in the same manner." (Anon. 1892).

For the year 1902 there is passing reference to an increase in production of dogfish oil by the Japanese oil and guano (fertilizer) factory at Departure Bay in the Nanaimo area, for use by the coal mining industry. In statistics for 1906 it was noted that there were two oil factories in District 1 (Vancouver and vicinity), but these were devoted entirely to oil recovery from salmon offal (Sword 1905); two in District 2 (north coast) of which one would have been at Skidegate; and three in District 3 (Vancouver Island): Departure Bay, Sidney (near Victoria) and one which cannot be traced but may have been in the Union Bay–Cumberland (coal mining) area, or to the south on one of the Gulf Islands or in the vicinity of Cowichan Bay (lumber industry).

From the beginning of coal mining near the mid-1800s until well after 1900, miners' lamps of the naked-flame type fuelled with dogfish liver oil were in extensive use on Vancouver Island (Fig. 19). We may hazard a guess at the amount of dogfish oil used in Vancouver Island mines, by using the recollection of one old miner that each man used a gallon of oil per month, for which he paid 35¢ (Bowen 1982:49). The annual report of the Department of Mines for British Columbia in 1892, records 2854 coal-mine employees on Vancouver Island (Anon. 1893). Thus, at 12 gallons per year per miner we arrive at a figure of 34 000 gallons or ($4.54 \times 35\ 000 = 155\ 500$ L). The

Fisheries and Oceans Canada | Pêches et Océans Canada

Canada

Common B.C. Groundfish

There are a variety of groundfish species on British Columbia's coast. Below are the most common groundfish encountered by recreational anglers. Remember: there is a rockfish conservation strategy in place to protect low numbers of inshore rockfish.

For more information on the conservation strategy, recreational fishing or fishing regulations, visit www.pac.dfo-mpo.gc.ca/recfish

Yelloweye Rockfish

Copper Rockfish

Canary Rockfish

Tiger Rockfish

China Rockfish

Quillback Rockfish

Bocaccio Rockfish

Redbanded Rockfish

Black Rockfish

Dusky Rockfish

Widow Rockfish

Yellowtail Rockfish

Kelp Greenling

Sablefish

Pacific Cod

Lingcod

Pacific Halibut

Arrowtooth Flounder

Starry Flounder

Rock Sole

English Sole

Pacific Sanddab

Longnose Skate

Big Skate

Ratfish

Spiny Dogfish

Red Irish Lord

Cabezon

Photos courtesy of:
1. M. Gjernes; 2. Hawkshaw

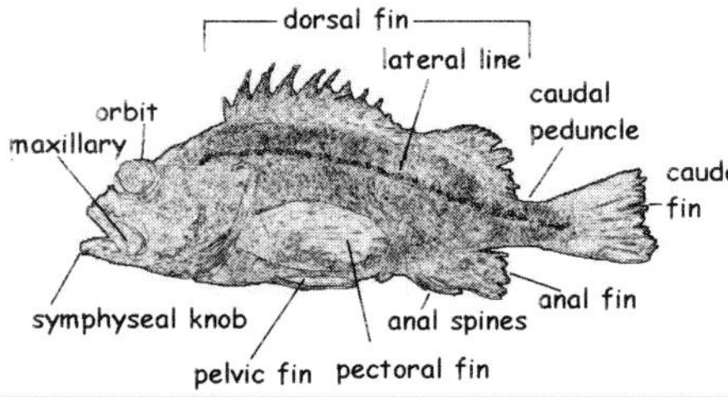

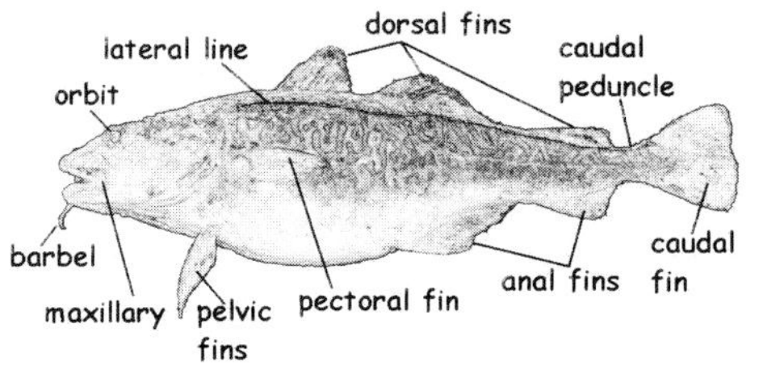

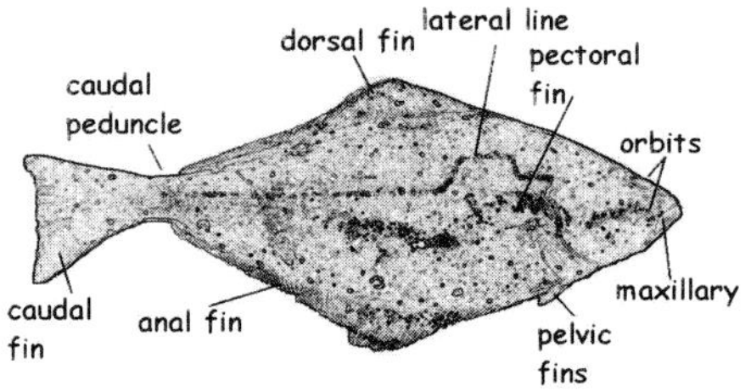

Tiger Rockfish
- Up to 61 cm
- Body white to red
- 5 dark red or black narrow vertical bands
- 2 dark red or black bands radiating from eyes
- Body can be brownish-red with black vertical bands

Canary Rockfish
- Up to 76 cm
- Body mottled orange to yellow on grey background
- Lateral line is pale
- Fins are bright orange
- 3 orange bands radiating from eyes
- Anal fin edge slants anteriorly

Copper Rockfish
- Up to 57 cm
- Body olive-brown to copper with pink or yellow blotches
- Body can be dark brown
- 2 yellow or dark bands radiating from eyes
- Last ⅔ of lateral line is pale
- Belly is pale pink to white

Yelloweye Rockfish
- Up to 100 cm
- Body yellow to red
- Eyes are bright yellow
- Fins usually have black tips
- Adults have light band on lateral line
- Juveniles are red with 2 light bands, one on lateral line and a shorter one below

Redbanded Rockfish
- Up to 64cm
- Body light pink to red with 4 broad vertical red bands
- 1 red band radiating from eyes

Bocaccio Rockfish
- Up to 91 cm
- Body dark orange-red to olive brown
- Lower jaw is long and projects past upper jaw

Quillback Rockfish
- Up to 61 cm
- Body dark brown to black mottled with orange-yellow
- Appears to have freckles
- Dorsal fin is high and deeply incised

China Rockfish
- Up to 43 cm
- Body black mottled with yellow, white and pale blue
- Broad yellow stripe starting at third dorsal spine and running along lateral line

Yellowtail Rockfish
- Up to 66 cm
- Body olive-green to green-brown
- Symphyseal knob present
- Anal fin edge almost vertical
- Fins have yellow tinge
- Jaw extends to back edge of orbit

Widow Rockfish
- Up to 53 cm
- Body golden-brown to light-brown
- Symphyseal knob absent
- Anal fin edge slants posteriorly
- Mouth is small
- Jaw extends to mid-orbit

Dusky Rockfish
- Up to 53 cm
- Body grey to greenish-brown fading to light grey or pink on belly
- Symphyseal knob present
- Anal fin edge almost vertical
- Brown bands radiating from eyes
- Jaw extends to end of pupil

Black Rockfish
- Up to 63 cm
- Body black to grey
- Belly is pale pink to white
- Fins are dark with black spots
- Anal fin edge rounded and slants anteriorly
- Jaw extends past orbit

Lingcod
- Up to 150 cm
- Body mottled brown to grey fading to white on belly
- Head, mouth and teeth are all large
- Appears to have 1 dorsal fin
- No barbel under chin

Pacific Cod
- Up 120 cm
- Body mottled grey to brown fading to white on belly
- 3 dorsal fins
- 2 anal fins
- Barbel under chin

Sablefish (Blackcod)
- Up to 107 cm
- Body black to grey
- Scales small
- 2 dorsal fins
- 1 anal fin
- Forehead flat
- Caudal fin forked
- No barbel under chin

Kelp Greenling
- Up to 61 cm
- Male body brown to olive with blue spots
- Female body light brown to golden blue with large brown to orange spots
- 5 lateral lines on each side

Rock Sole
- Up to 60 cm
- Body mottled brown
- Dark blotches on fins
- Blind side white with pink tinge
- Mouth small
- Scales large and rough
- High arch on lateral line
- Right-eyed

Starry Flounder
- Up to 1 meter
- Body brown to green and diamond shaped
- Blind side white to tan
- Dorsal and anal fins are banded with black
- Scales rough
- Can be right or left-eyed

Arrowtooth Flounder
- Up to 84 cm
- Body brown-grey to olive
- Blind side white to grey
- Mouth large
- 2 rows of large arrow-shaped teeth
- Caudal fin forked
- Right-eyed

Pacific Halibut
- Up to 270 cm
- Body marbled brown with grey
- Blind side white
- Body thick and sturdy
- Mouth large with sharp conical teeth
- Caudal fin slightly forked
- Almost always right-eyed

Big Skate
- Up to 240 cm
- Body olive-brown to grey
- Blind side is white
- Dark eye spots on wings
- 5 gill slits
- Dorsal spines start above the tail

Longnose Skate
- Up to 140 cm
- Body dark brown
- Blind side is grey
- Long pointed nose
- 5 gill slits
- Dorsal spines start at tail

Pacific Sanddab
- Up to 41 cm
- Body brown to tan mottled
- Blind side white to tan
- Caudal fin rounded
- Eyes and mouth are large
- Left-eyed

English Sole
- Up to 57 cm
- Body light brown
- Blind side white to yellow
- Body smooth and diamond shaped
- Head and jaw pointed
- Right-eyed

Cabezon
- Up to 100 cm
- Body marbled olive-green to brown-grey with white patches
- Body can be red
- Flap-like projections on snout and over each eye

Red Irish Lord
- Up to 51cm
- Body red mottled with brown, white and black
- 4 vertical dark bands
- Single dorsal fin notched to form 3 steps
- Snout blunt and rounded

Spiny Dogfish
- Up to 160 cm
- Body slate grey to brown
- Belly white to light grey
- 5 gill slits
- 2 dorsal fins with a spine in front of each
- No anal fin

Ratfish
- Up to 100 cm
- Body grey-brown with white spots with olive belly
- Tail is long and tapering
- Watch out for a poisonous spine at the front of the dorsal fin

For reference purposes only.

More detailed information on these and other groundfish species is available by consulting a fish identification publication.

FIG 19. Coal miners from the Extension mine near Nanaimo (ca. 1900) wearing caps equipped with open-flame pitlamps. Photograph courtesy of the Provincial Archives of British Columbia (Cat. No. 80599). *Inset*: A miner's dogfish oil lamp. The lamp was filled by opening a hinged lid on top; a cotton wick was inserted down the spout and then lighted (photo by author from collection in the Nanaimo Centennial Museum).

equivalent round weight of dogfish is about 1870 t. This is an overstatement to the extent that not all mine employees worked underground with pit lamps, but on the other hand it could be an understatement since it excludes amounts of oil used for lubrication and other lighting needs. The figure does not seem unreasonable, for we have already recorded (Table 3) minimum and maximum estimates for the Strait of Georgia in 1892 of 1435 and 2464 t, respectively.

Although dogfish-oil lamps were gradually replaced by safety lamps, a government commission as late as 1902 continued to endorse their use as long as the coal companies observed other safety precautions in the mines. However, ". . . it was not until 1918 after sixteen more years of dreadful explosions and loss of life that naked lights were eliminated from all Vancouver Island coal mines with the exception of the Nanoose Collieries Mine at Lantzville" (Bowen 1982: 117).

It is informative to note the kinds of printable language used in discussion of dogfish when the species is of little or no economic value, as was to happen often in British Columbia. It had been a problem of much longer standing in eastern Canada, where the Inspector of Fisheries from New Brunswick, in his 1904 report to the Dominion Commissioner of Fisheries, stated the following:

> "The ravages of these scavengers of the sea have been written about so frequently to your department during the past few seasons. . . that it is needless for me to refer at any length to this important subject. . . What action to take on these sea wolves is a subject of serious concern for the whole North Atlantic seaboard, and it is earnestly hoped that vigorous steps will be taken which will lessen the ravages of the voracious fish, or that the schools of dogfish will make one of those surprising and mysterious movements with which they are credited and disappear from our coasts with the same rapidity that they invaded them." (Pratt 1904).

In the early autumn of 1907 a fully equipped whaling station was established at Pages (now Piper's) Lagoon at the entrance to Departure Bay near Nanaimo, but before the end of the year disappointing catches of whales in the Strait of Georgia had necessitated acceptance of dogfish for conversion to oil and fertilizer (Taylor 1908). However, with declining markets, this substitution was insufficient to make ends meet. By 1908 the venture had been terminated and the plant was dismantled and moved to the Queen Charlotte Islands.

According to Lyons (1969) the Skidegate Oil Co. was bought by Simon and Leiser & Co. of Victoria in the late 1800s, and was operated (although latterly not every year) until 1912 when it was resold to British Columbia Fisheries Ltd. This discontinuity is not apparent in Fig. 16, but a long-term decline was in progress. If the historical statistics are accurate, there may have been more than one oilery in operation on the Queen Charlotte Islands.

During the early 1900s there were two dogfish reduction plants in Southeast Alaska, one called the Revilla Reduction Works located at Ketchikan and the other, the floating plant *Elliot* operated by W. H. Royden Ltd. of Petersburg. The former stopped operating in 1911 because the supply of dogfish was disappointing while the latter appeared to terminate operations in 1915 (Pacific Fisherman Yearbooks 1912–16).

Certainly off the south coast severe economic difficulties are suggested by Fig. 16 in the collapse of the fishery in 1910. Production of oil in the north, although at a much lower level than in the 1880s, did not collapse completely until the early years of World War I.

The decline in demand for dogfish was due in part to the introduction of calcium carbide lighting in local industries and the displacement of dogfish-oil lamps in the mines by safety lamps and eventually (1917) by electric lamps. The bitter industrial strike by coal-miners on Vancouver Island from 1912 to 1914 probably contributed to this decline. Furthermore lubricants derived from petroleum may have become competitive once rail links were established with the manufacturing centers of eastern Canada.

In any event, the let-up in fishing pressure in the Strait of Georgia prompted Taylor (1916) to remark that "...the dogfish are another, and considered by many, even a more serious menace to the fisheries than the hair seal and sea-lion, as they not only destroy valuable food fish but work havoc with the fishermen's net... and have evidently become more numerous during the past few years." He stressed the need for re-development of a commercial fishery for dogfish.

Towards the end of World War I there was an acute shortage of meat in the United States which may have been felt as early as 1916 for in that year there was a Canadian sale of 250 t of dogfish to the San Juan Fishing Co. at Friday Harbor (in U.S. waters of the Strait of Georgia) (Taylor 1917). The catch was canned but the product proved to be unacceptable because of an in-the-can breakdown of urea into ammonia and carbon dioxide (Alverson and Stansby 1963).

In regard to other operations in the State of Washington, it is apparent that the four to five reduction plants located in Puget Sound were used primarily for the rendering of salmon offal to oil and meal. However, a few plants specializing in dogfish were reported in Pacific Fisherman Yearbooks, such as the Pacific Products Co. at Port Townsend which started operating in 1914. Yet these establishments were to play but a minor role in events to come.

3. The Industrial Oil and Meal Era (1917–39)

Beginning in 1917, published fishery statistics for British Columbia once again made provision for spiny dogfish in landing records (Table 2). They indicated the amounts of fish landed but not until 1930 would they provide additional information on marketed products. Operations were now confined to the south coast District 3 and within that to subdistricts representing the Strait of Georgia. Operations in the Queen Charlotte Islands apparently ceased after 1916 and would not resume until nearly 25 years later. (For comments on the interpretation of the 1917–39 records which forms the basis for Fig. 20 and Table 4, the reader is referred to Appendix 3, section 3.)

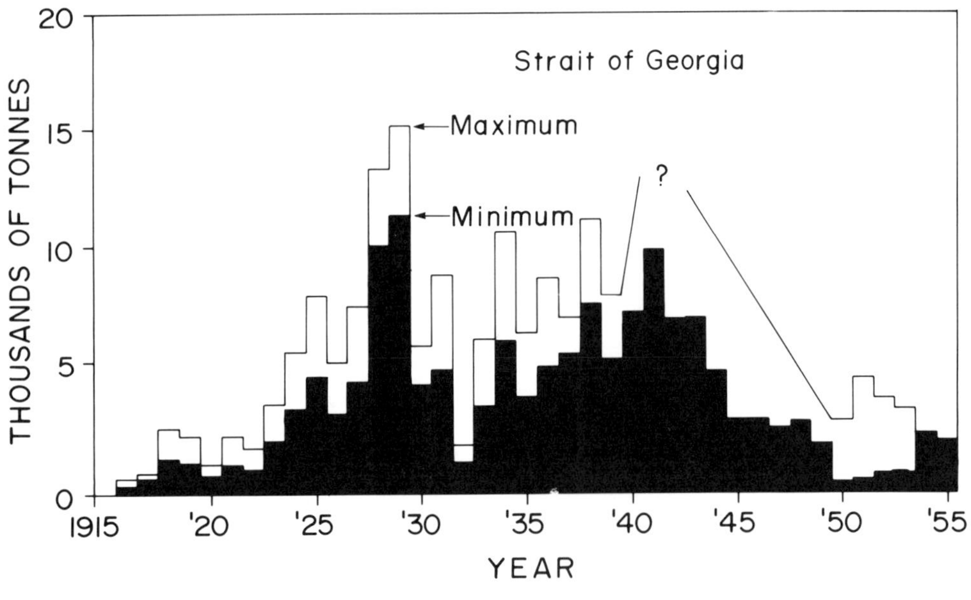

FIG 20. The 1916–55 catch of spiny dogfish in the Strait of Georgia as estimated from actual commercial landings and records of oil and meal production.

TABLE 4. Estimated minimum and maximum total weights (tonnes) of spiny dogfish caught in the Queen Charlotte Island portion of District 2 and in the Strait of Georgia portion of District 3 compared with totals for all three Districts of B.C. Data derived from Appendix 3.

Year	Queen Charlotte Islands		Strait of Georgia		British Columbia total	
	Min	Max	Min	Max	Min	Max
1917	—	—	508	805	508	805
18	—	—	2685		2685	
19	—	—	2310		2310	
1920	—	—	680	1190	680	1190
21	—	—	2384		2384	
22	—	—	1825		1825	
23	—	—	2206	3714	2206	3714
24	—	—	3357	5938	3357	5938
1925	—	—	3358	8438	3358	8438
26	—	—	3555	5505	3555	5505
27	—	—	5112	7945	5112	7945
28	—	—	10452	13907	10458[a]	13907
29	—	—	11773	15866	13279[b]	15866
1930	—	—	4476	6262	4476	6262
31	—	—	5096	9303	5096	9303
32	-	—	1271	1923	1271	1923
33	—	—	3611	6419	3611	6419
34	—	—	5308	11143	5308	11143
1935	—	—	3484	6683	3484	6683
36	—	—	5268	8991	5268	8991
37	238	409	5136	6803	5374	7212
38	231	397	7243	10794	8170[c]	11887
39	56	96	4498	7212	5208	11862

[a]Includes 6 t reported from areas other than Q.C.I. in District 2.
[b]Includes 1506 t reported from areas other than Q.C.I. in District 2.
[c]Includes 596 t reported from west coast of Vancouver Island portion of District 3.

The market for dogfish as food for export to the United States lasted through 1918 with landings originating from Canadian waters of the Strait of Georgia reaching 2685 t in that year. The fishery failed to live up to expectations as more traditional species became available again after the end of World War I.

However, interest quickly turned to manufacture of meal for cattle and poultry food, the rendering of liver oil for use in poultry and medicinal preparations, as a base for insecticide sprays and for the tanning of leather. Body oil and lower grades of liver oil had such varied uses as in the tempering of steel and in the manufacture of sheep and cattle dips (Brocklesby 1941).

Although production was at a much lower level from 1918 to 1922 than in the early 1900s, there were seven plants in operation in the Strait of Georgia. Pacific Fisherman Yearbooks (1921–22) note plants at Nanaimo (Nanaimo Fish Products), on Parker Island in Trincomali Channel (Veterans Products Ltd.) and one near the entrance to Bute Inlet (Rendezvous Fisheries Ltd.) which was later (1924) to move to Lasqueti Island and become the False Bay Fisheries Ltd. At one of these an attempt was made to produce leather from dogfish skin but this was short-lived because of the state of the economy (Motherwell 1922). During the 1920s the Francis Millerd Fishing Co. acquired the former Union Steamship Co. freighter M/V *Chilliwack* which was converted to a floating dogfish reduction plant and renamed the *Graylor*. No records can be found of its area and duration of operation other than its association with a salmon cannery at Sointula near Alert Bay where dogfish oil was manufactured and shipped to Vancouver (*Pacific Fisherman*, Sept. 1927). It is evident from correspondence, however, that the Department of Fisheries was reluctant to issue a licence for mobile operations when the company applied in 1927 for permission to operate in Queen Charlotte Strait (F. W. Millerd, personal communication). By 1928 the *Graylor* was being used as a power plant at a pilchard reduction plant in Kyuquot Sound (*Pacific Fisherman*, Sept. 1928).

In 1922 markets for fishmeal, fertilizer, and oil began improving, but the industry ran into allegedly unfair competition with the United States. Motherwell (1923) reported that

> "...in waters adjacent to the American boundary in the south, as a result of the protection afforded American fishermen by the Fordney–McCumber tariff, it has become possible for American buyers to come across from Puget Sound and outbid the Canadian establishments by as much as $3 per ton for the raw products in the way of grayfish (dogfish) captured on this side of the line."

Although there were a few plants in Puget Sound that entered the dogfish oil and meal business (the Puget Sound Reduction Co. at Blaine and another at Anacortes) it is evident that most of the raw material was coming in from British Columbia. Operators there "... have been quite successful in handling dogfish and believe that they have learned methods of fishing [with sunken gill-nets] that will assure a supply of fish" (Pacific Fisherman Yearbook, 1923). However, in 1923, the Puget Sound dogfish industry was set back by a Canadian law prohibiting the export of dogfish to U.S. reduction plants, possibly in response to Canadian industry complaints about unfair practices the previous year. To 1923 at least it had not been found possible to secure a regular supply of dogfish from Puget Sound waters (Pacific Fisherman Yearbook, 1924).

After 1922 the landings of dogfish began to increase rapidly due to greatly strengthened markets for fish oil and meal generally. The increase was much more dramatic in other Canadian fisheries. Permission to reduce Pacific sardines (pilchards) *Sardinops sagax* and herring in the northern part of British Columbia was granted in 1925 but no activity in respect to the latter species occurred until permission was extended the following year to waters off the northeast coast of Vancouver Island. The number of licences issued rose sharply from seven to 23. Of this number 21 were issued for Vancouver Island (west coast, because this was where the sardine fishery took place), but only 14 plants were actually constructed.

Landings of dogfish increased from about 2500 t in 1921–23 to at least 13 000 t in 1929 and the number of "grayfish" fishing licences issued increased from about 100 to 420 (Table 5). During the 1922–29 period the principal method of fishing consisting of baited set-lines

TABLE 5. Numbers of spiny dogfish licences issued to British Columbia fishermen according to racial origin[a].

Year	Whites	Indians	Others[b]	Total
1922	31	—	89	120
23	26	3	58	87
24	54	—	143	197
1925	68	5	92	165
26	79	6	92	177
27	137	7	93	237
28	219	11	191	421
29	149	35	238	422
1930	90	2	228	320
31	32	—	213	245
32	9	—	123	132
33	9	—	65	74
34	31	1	120	152
1935	23	1	88	112
36	43	—	81	124
37	56	11	94	161
38	169	20	299	488
39	95	12	258	365
1940	164	2	239	405
41	357	121	419	897
42	1026	208	—	1234
43	1807	242	—	2049
44	2745	320	—	3065
1945	2031	135	—	2166

[a]Source: Annual Reports of the Western Division of the (Canadian) Fisheries Branch, of the Department of Marine and Fisheries (1922–29), and of the Department of Fisheries (1930–45).

[b]Almost exclusively fishermen of Japanese extraction.

TABLE 6. Numbers of spiny dogfish licences issued to British Columbia fishermen, by type of gear, for selected years.[a]

	Type of gear		
Year	Hook and line	Sunken gill net	Total
1923	47	40	87
1924	155	42	197
1925	133	32	164
1926	?	?	177
1927	190	47	237
1928	289	130	419

[a]Source: Annual Reports of the Fisheries Branch of the Department of Marine and Fisheries. For years other than those shown no distinction was made as to type of gear.

strung out on the ocean bottom and sunken gill-nets made of discarded sockeye netting (Table 6). It is known that there were a few otter-trawlers or bottom draggers operating in those years but it is uncertain whether they produced significant landings of dogfish.

In 1927 a dogfish reduction plant was established by Pender Island Fish Products Ltd. at Shingle Bay on North Pender Island (in Minor Statistical Area 18 in Fig. 3). This plant had three batch cookers which could handle salmon offal from Vancouver and Vancouver Island canneries in summer months and dogfish during the winter season (*Pacific Fisherman*, May 1927). In addition to the plant at Nanaimo which had been in existence since the early 1920s, another appeared on the scene in 1931 at Deep Bay (Fig. 3) as Sea Fisheries Ltd., adjacent to a salmon cannery run by the Deep Bay Packing Co. (*Pacific Fisherman*, Feb. 1931). This operation, later taken over by B.C. Packers, was likewise designed to use salmon offal in summer and dogfish in winter.

At Tacoma in southern Puget Sound a floating reduction plant, the schooner *Meteor* (owned by Marine Products Corporation), began operating on dogfish in the winter of 1926–27 (*Pacific Fisherman*, Mar. 1927). This appeared to be a short-lived venture, for the vessel was designed primarily for reduction of salmon offal during summer-time operations in Alaska.

After World War I special effort was made to find employment in British Columbia fisheries for veterans. The federal government took steps to eliminate all orientals from West Coast fisheries by 1927. Largest reductions were effected in the salmon and herring fisheries, but in 1926 the Department of Marine and Fisheries acceeded to a request from British Columbia reduction plant operators and granted Japanese fishermen licences to catch dogfish with hand-lines (*Pacific Fisherman*, July 1926). Operators had experienced considerable difficulty in stimulating interest among white fishermen. In most years through the 1920s and 1930s the majority of fishermen engaged in the fishery for dogfish were of Japanese extraction and most were Canadian citizens. Their predominance came to an abrupt end in 1942 when they were stripped of their vessels, homes and other possessions, and evacuated from coastal areas of British Columbia because of the war in the Pacific (Table 5).

As an effect of the economic "crash" of 1929, production fell sharply in 1930 and still further through 1932, to little more than 1200 t. Prices paid for round dogfish, which had risen from $5.00 to $7.00 per ton between 1922 and 1929, fell to $3.00 in 1932 (Table 8). Slow recovery in price and amount landed began immediately thereafter but not before several reduction plants were forced into bankruptcy, including the Nanaimo Reduction Works located in the Newcastle Channel portion of Nanaimo harbor. The plant at Shingle Bay was still in operation, but it too had run into financial difficulties as early as 1928 and had been taken over by North Pender Island Reduction Works Ltd. (*Pacific Fisherman*, Dec. 1928).

Reduction of dogfish to fish meal was a special process different from that used for herring and Pacific sardines. Whole fish were placed in a steam-jacketed cauldron and cooked under pressure while being stirred by revolving paddles. If livered carcasses were used (as in the late 1930s and early 1940s) they were cooked in salmon oil. When flesh and skin had disintegrated, the mixture was cooled and spread between two layers of burlap. A press, not unlike those used in making plywood, was then used to extract the liquid and compress the flesh into a cake or sheet measuring roughly 120 cm square and 2–4 cm thick. When dried, the cake was broken in pieces and put through a grinder to produce fish meal. Occasionally it was left as broken pieces for shipment to China as food. The extracted mixture of dogfish liver and body oils (latterly body oil only) and salmon oil was a byproduct of this operation (D. R. Russell, Vancouver, and H. Mosdell, Nanaimo, former reduction plant operators, personal communications).

By the late 1930s in the Strait of Georgia there were only two plants that remained involved in the reduction of dogfish, the one at Deep Bay at the southern end of Baynes Sound and the other at Shingle Bay on North Pender Island. A third is mentioned in Department of Fisheries reports, but its identity remains a mystery. It may have existed only on paper as an issued but unutilized licence. Regrettably, many details of the operation of these reduction plants have now been lost. Fisheries Statistics of Canada failed to distinguish among the various kinds of reduction plants until 1943, when the three noted above were first mentioned (Table 7).

To this point we have said nothing of developments off the north coast of British Columbia. If we were to believe published statistics, it would appear that a fishery was nonexistent for more than 20 yr after its collapse in 1916 (Tables 3 and 4). However, from other sources there are indications of some activity in the late 1920s. The Prince Rupert Marine Products Co. made plans in 1927 to enter the dogfish reduction business to keep its plant (which normally handled nothing but salmon offal) busy during the winter months. An 80-foot tender the *Red Boy* was purchased from the Skeena River Packing Co. Ltd. and used in the autumn of 1928 to haul at least one load (70 t) of dogfish to Prince Rupert from Skidegate Inlet (*Pacific Fisherman*, Nov. 1928). From what we know today about the seasonal availability of dogfish in Hecate Strait, to say nothing of weather conditions, a winter operation dependent on grounds near the Queen Charlottes was doomed to failure. By the winter of 1930–31 the Prince Rupert company had ceased dogfish reduction operations (*Pacific Fisherman*, Mar. 1931). Although yield from Hecate Strait during 1927–30 was probably of low magnitude, no records of actual production can be found. Rewakening of the Hecate Strait fishery was nearly another decade away.

TABLE 7. Numbers of Canadian plants operating to reduce spiny dogfish carcasses to fishmeal and plants to convert fish liver to oil and vitamin A[a].

Year	Dogfish reduction plants	Fish liver reduction plants[b]
1941	3	?
42	3	9
43	3	8
44	3	8
1945	3[c]	10
46	2	7
47	1	7
48	—	6
49	—	6
1950	—	4
51	—	4
52	—	4

[a]Sources: 6th to 8th Annual Report of the Federal Department of Fisheries; Annual Reports of the Provincial (B.C.) Fisheries Department. Prior to 1941 no distinction was made between the numbers of dogfish reduction plants and those reducing offal from other species including salmon.

[b]These plants processed dogfish liver predominantly, but smaller quantities of livers from soupfin shark, halibut, lingcod, etc. were also used.

[c]Although licences continued to be issued (or to be reported as issued), no plants were operating from 1945 onwards.

TABLE 8. Average price (per ton) of whole spiny dogfish and per pound of liver landed in British Columbia.[a]

Year	Whole fish	Liver	Year	Whole fish[b]	Liver
1917	$ 9.95	—	1940	$ 8.10	$0.055
18	10.00	—	41	14.10	0.094
19	7.00	—	42	2.40	0.162
1920	6.00	—	43	2.50	0.263
21	4.95	—	44	2.20	0.343
22	5.00	—	1945	—	0.315
23	5.25	—	46	—	0.312
24	6.25	—	47	—	0.289
1925	6.00	—	48	—	0.370
26	6.00	—	49	—	0.300
27	7.00	—	1950	—	0.243
28	7.00	—	51	—	0.268
29	7.00	—	52	—	0.222
1930	6.15	—	53	—	
31	4.10	—	54	—	0.163
32	3.00	—	1955	—	0.156
33	3.45	—	56	—	0.118
34	3.75	—	57	—	0.122
1935	4.00	—	58	—	0.103
36	4.55	—	59	—	0.155
37	5.75	0.060	1960	—	0.147
38	3.40	0.053	61	—	0.139
39	4.15	0.098			

[a]Source: Dominion Bureau of Statistics, Fishery Statistics for the Province of British Columbia.

[b]1942–44, Price for livered carcasses only.

Through 1935, published statistics suggested that the reduction plants in the Strait of Georgia were processing whole dogfish and extracting a mixture of both liver and body oil, but by 1937 they were separating livers from bodies and marketing the livers separately. The value derived from livers was reflected in the rise in price paid for whole dogfish (Table 8). In 1937, 82 t of liver were shipped to the United States for production of a vitamin A oil additive to poultry food (Brocklesby 1941). This was in response to developments in California which marked the beginning of a new and most dramatic "Cinderella" phase in the long and generally lack-lustre history of the fishery for spiny dogfish.

4. The Great Vitamin Liver Fishery (1937-50)

a) Discovery of Vitamin A Content

To appreciate new and spectacular events about to take place in the dogfish fishery, it is instructive to retrace our steps to 1926. In the summer of that year the Biological Board of Canada established its Pacific Coast Technological Station in the seaport town of Prince Rupert, B.C., to provide for study of the chemistry and bacteriology of fish products and by-products, as well as problems in the preservation of such products. One of the staff was H. N. Brocklesby whose specialty was the study of fish oils and meal, particularly that of dogfish. Although many attempts had been made to utilize dogfish, the production of fish meal and oil had never been very successful because of inadequacies in the methods used. Brocklesby found that dogfish liver oil possessed properties that would make it valuable in leather and tanning industries, in manufacture of certain paints, soaps for various purposes and as a food additive. More important, as future events would prove, was his discovery that the vitamin A potency of dogfish liver oil was five to ten times greater than that in standard medicinal cod liver oil (Brocklesby 1927). The actual discovery that dogfish liver oil was a good source of vitamin A apparently was made by Holmes and Pigott (1925), quoted by Pugsley (1940), but it remained for Brocklesby to demonstrate *how* good.

Brocklesby went on to find that the vitamin D content of dogfish oil was only one tenth to one third that of cod liver oil. With others at the Prince Rupert Station, he advocated the utilization of dogfish liver oil for fortifying the vitamin A content of certain British Columbia fish oils containing vitamin D to produce blended oils having the ratios of vitamin A and D desired by stock and poultry feeders (Swain 1944). The recommended blending of dogfish and other oils, after further experimentation, eventually received wide acceptance by the livestock and poultry industries.

Interest in dogfish liver oil in the USA began in California between 1936 and 1937 (Harrison and Sampson 1942a). At that time a search was being made by sardine (pilchard) processors for a cheap source of vitamin A with which to fortify sardine oils being offered in the animal-feed market. This provided the initial stimulus for development of a fishery specifically for dogfish in Washington, Oregon, and California. Canadian companies already in the dogfish business began shipping liver to the United States market. By 1938, the average price offered by Seattle buyers (representing California interests) was 5¢/lb, or about the same as that offered to Canadian reduction plant operators and to the relatively few fishermen who were extracting livers and discarded carcasses at sea.

Activity in the U.S. northwest states suffered a temporary setback early in 1939 when California producers discovered that liver of the soupfin shark *(Galeorhinus zyopterus)* was practically as fat as that of dogfish, vitamin potency of the oil was considerably higher, and soupfin sharks could be obtained in sufficient quantity to meet their then relatively moderate demands. This interest in soupfin continued through 1940, all along the Pacific coast but there was also a renewed demand for dogfish liver.

b) Rising War-Time Demand

Harrison and Sampson (1942a) listed a number of factors that contributed to this resurgence:

"1. The outbreak of war in Europe in 1939 and the invasion of Norway in the spring of 1940 cut off the normally large imports of cod liver oil, thus increasing the demand for vitamin oils of domestic origin.
2. An economical process was developed for concentrating vitamin A from oils of relatively low vitamin potency, such as dogfish liver oils.
3. The relatively low price at which (soupfin) shark and dogfish liver oil vitamin A could be prepared made this source increasingly competitive with the established vitamin liver oils and created a greater incentive for fortifying foodstuffs.
4. Recognition of the potentially greater market (for soupfin shark and dogfish liver oils) caused new oil producers to enter the field in all three coastal states, thus increasing the competition for raw materials."

Harrison and Sampson went on to say that

"...the year 1941 will long be remembered in the annals of the Pacific shark industry. Offerings for the catch began to spiral (in the latter half of the year)... In Oregon and Washington the price for... dogfish livers rose from 8-10 cents per lb. to over 50 cents per lb."

Prices quoted were at best half-year averages and may have represented just special purchases. In any event, initial developments in Canada appeared to be more modest. At Vancouver the average annual price paid for dogfish liver stood at 6¢ in 1940, then rose to 9.4¢ in 1941 and 16.2¢ in 1942 (Table 8). Canada was supplying domestic and British markets as well as those in the States and prices were less affected by developments in the latter country: viz., a shutting off of Japanese imports and a move by the U.S. government to buy large lots of vitamin A for export to friendly nations under the Lend–Lease Act.

Vancouver prices rose erratically after 1941 to 38¢/lb by late 1943, and Seattle prices went as high as 67¢/lb. In the United States, however, the 1941–43 prices were very unstable because of the speculative nature of buying (in anticipation that the Food and Drug Administration would lift the price ceiling). The FDA and producers agreed on a scheme which by late 1943 had a stabilizing effect on prices, namely a "set-aside" program requiring each producer to sell to the government a fixed percentage of his available stocks before they could be disposed of on the open market (Pacific Fisherman Yearbook, 1944).

In Canada, through the Agency of the Canadian Oils and Fats Administration, 30% of the vitamin A produced had to be retained for domestic consumption. The remainder found a ready market in the U.K. and USA.

To those Canadian fishermen landing round dogfish in the Strait of Georgia, the price rose from $4 per ton in 1939 to $14 in 1941. Initially, reduction plant crews removed the livers for resale and then reduced the remaining carcasses. The same occurred in the United States but as the value of livers increased fishermen became suspicious of the "rule of thumb" that livers represented 10% of the whole fish weight. By 1940–41 in California, fishermen were themselves gutting their dogfish catches, selling the livers and discarding at sea whatever carcasses could not be marketed ashore (Anon. 1949). The same change in practice took place in Washington and Oregon. In British Columbia fewer and fewer round dogfish and livered carcasses were delivered to the Strait of Georgia reduction plants until, by the summer of 1944, both had ceased to operate, although licenses continued to be issued until 1947 (Table 7). R. E. Walker, production manager of B.C. Packers, in announcing closure of the Deep Bay

TABLE 9. Illustration of the rapid transition in practices of landing and marketing of spiny dogfish in British Columbia, 1939–45 (weights in tonnes, except for oils in thousands of litres).[a]

Year	Dogfish caught and landed	Dogfish livers landed	Marketed			
			Liver	Liver oil	Body oil	Meal
1939	5208	—	27	591	—	839
1940	6412	711	267	292	432	981
1941	6486	1578	79	965	328	953
1942	4560	1924	33	1395	191	829
1943	3580	2072	25	1747	168	642
1944	1104	3524	82	2447	82	299
1945	—	2641	—	1959	—	—

[a]Source: Fisheries Statistics of Canada.

Note: The length of this transition is probably greater than that shown here, because keepers of a statistical system are unlikely to respond to the need for a change of format until a trend in purchasing practice is well established. For example, Capt. Jim Pope (retired fisherman of long and varied experience) advises that, at his Newcastle Island base near Nanaimo, he began purchasing dogfish liver directly from gillnet fishermen in 1935. Yet clear evidence of such practice did not appear in official statistics until 1940.

plant, said that ". . . the government had declined to yield to a suggestion that fishermen be required to retain the carcasses for marketing" (*Western Fisheries*, Sept. 1944).

In 1942, the first year in which the Provincial Government issued licences, there were nine plants processing liver into oil and into vitamin A. By 1945, 10 plants were operating (Table 7). The trend in Canada towards landing of livers only and discarding of carcasses at sea and the trend away from marketing (exporting) of unprocessed liver as Canadian companies were established to produce vitamin A, are illustrated in the series of statistics for 1939–45 shown in Table 9. Further changes, particularly in purchasing practice, were still to come.

c) The Fishing Fleet

It is necessary to describe briefly the evolution of the various kinds of fishing vessels used to catch dogfish as Canadian and United States fishermen succumbed to the "gold rush" fever.

i) Set-line (longline) vessels — Up and down the Pacific coast prior to 1939 the traditional method of catching dogfish was by baited set-line, which consisted of a main line hundreds of metres in length to which branch lines (gangions) were attached at 1–3 m intervals to which in turn hooks were attached. The gear involved a relatively small outlay of capital and, more important in years to come, it tended to catch mainly the larger sizes of fish. In Canada, in addition to the small vessels involved in the Strait of Georgia, larger offshore vessels used for halibut began to enter the scene because the legal length of the halibut fishing season in British Columbia waters was becoming progressively shorter. By 1939 it was down to 120 days and by 1944 it was only 51 days and still falling (Anon. 1948), thus providing more and more time for fishermen manning these vessels to supplement their incomes. United States halibut vessels based in Seattle entered the soupfin and dogfish set-line fishery as far south as San Francisco (Harrison and Sampson 1942a).

ii) Sunken gillnet (set-net) vessels — The use of gill nets specifically for both soupfin shark and dogfish came into vogue in California in 1940–41 (Anon. 1946) and in Washington and Oregon in 1941–42. In the latter states, fishermen initially resorted to the use of discarded salmon gill nets. This gear, though tending to be selective of large fish appeared to be less selective than longline gear.

Sunken nets (set on the ocean bottom) had been permitted in Canadian waters since the early 1920s, but it was not until 1939 when they became widely used in the Strait of Georgia that the first of a number of restrictions (an area closure) was imposed in response to representations from fishermen who were involved in the long-established hand-line fishery for lingcod *(Ophiodon elongatus)*. It was claimed that nets set on reefs, specifically the Gabriola Reef and nearby channels among the northern Gulf Islands, depleted the supply of lingcod. This regulation was followed by another in 1941 which prohibited sunken-net fishing anywhere in the Strait of Georgia from November through March. Finally, in July 1944 they were banned from all British Columbia waters, but within a matter of weeks the order was amended to apply only to the Strait of Georgia and a few minor localities elsewhere (Barraclough 1948a).

Meanwhile in open coastal waters the fishery for dogfish with sunken nets made of discarded salmon gill nets of cotton thread came into increasingly widespread use in 1942 — at about the same time that the power-driven drum was adopted on many gillnet vessels. This opened the way for extensive participation of small-boat operators on grounds adjacent to Barkley Sound on the west coast of Vancouver Island and especially on the Hecate Strait flats, the site of fishing during the early days (1876–1916) of the dogfish fishery. Between 1941 and 1945 the numbers of set-nets increased from 117 to 1671 (Table 10).

TABLE 10. Statistics on numbers of vessels, boats and kinds of gear in use in British Columbia with potential for involvement in the fishery for spiny dogfish[a].

Year	"Groundfish"[b]		"Small dragger" licences	Amount of gear		
	Vessels	Boats		Set-nets[c]	Small drag nets	Skates of gear[d]
1935	26	250	29	47	11	1469
36	31	266	31	28	14	1659
37	23	334	33	39	13	2483
38	26	324	43	32	15	1844
39	37	347	47	33	22	2777
1940	40	269	45	47	31	2814
41	65	366	45	117	35	3887
42	86	481	42	329	31	4674
43	51	786	43	305	42	7000
44	80	866	65	775	77	8136
1945	86	873	88	1671	68	8583
46	65	932	107	1345	71	10923
47	112	727	111	876	108	9658
48	188	510	92	697	80	8301
49	?	?	86	993	87	8163

[a]Source: Fisheries statistics of Canada.
[b]Presumably refers to vessels/boats fishing for halibut, lingcod and other groundfish in addition to spiny dogfish.
[c]Excludes gill nets used for salmon, herring, and smelts.
[d]Includes gear used for halibut or sablefish.

iii) Otter-trawl vessels — Since well before World War I, there had always been a few trawlers working in close proximity to ports of Vancouver and Prince Rupert supplying the very limited market demand for several species of flounders, Pacific cod and rockfish — fishes that were not readily available to other types of gear. Trawl nets are funnel- or bag-shaped nets that are dragged along the ocean bottom and kept open by a pair of otter-boards. The fish collect in the cod-end which is opened on deck when the net is brought aboard (Fig. 21). During the 1930s, between 23 and 47 vessels were licensed to fish with otter-trawls, a smaller number actually participated and even fewer fished year-round (Table 10). It is not likely that any of these vessels played a significant role in the Strait of Georgia fishery for spiny dogfish in those earlier years, but after 1937 this situation changed significantly. The numbers of licenses began to increase and by 1945 there were 107 part- or full-time vessels in operation.

The same phenomenon, but on a grander scale, occurred in Washington waters where at least until 1936 it appears that dogfish had no value and were discarded at sea (Smith 1936). The numbers of licensed trawlers increased from 43 in 1937 to 255 by 1945, and in Oregon from 4 to 83 (Table 11). This growth both in Canada and the United States, though stimulated primarily by the demand for dogfish liver, was aided by a more gradual increase in the demand for fish fillets to offset war-time shortages of meat — a repeat on a much larger scale of events in 1914–18.

Trawl nets, unlike set-line and sunken gillnet gear contributed to a growing problem concerning the marketing of dogfish liver, for the gear was relatively unselective for size of fish. Consequently, with demand running high, trawl fishermen landed livers from fish of a wide range of sizes and by 1944, at least in the State of Washington, the major share of the dogfish catch in both inshore and open-coast waters was being taken with trawling gear (Anon. 1944a). In addition the stocks were beginning to show the classical effects of overfishing: the supply of large sizes of dogfish available to *all* types of gear was starting to decline but was most evident in the set-line fishery (Holland 1957).

d) Buying Dogfish Liver "On Test"

Pugsley (1937) was the first to note the wide variation in vitamin potency of individual dogfish livers and in the limited sampling he conducted he found that pregnant (large) females and large males had livers containing the highest levels of vitamin A. Harrison and Sampson (1942b) noted individual livers ranged from 50 to 75% oil by weight and varied between 5 000 and 30 000 U.S.P. units of vitamin A per gram of oil but in 1942 most fell within the 10 000 to 20 000 unit range. Bonham et al. (1949) were later to show that the range could be as much as 1000 to 39 000 units per gram. Fish oil companies were quick to notice this wide variation in potency from vessel to vessel and even among liver cans or drums from the same vessel, and equally quick to see the disadvantage of paying a fixed price per pound of liver, which had been the practice since 1937. In 1944 the companies started buying liver on the basis of assayed oil content and vitamin potency and by the late spring of 1945 the practice had become universal in the United States (Pacific Fisherman Yearbook, 1946). To judge from the remarks of Hart (1946) "buying on test" was still not widespread in British Columbia even early in 1946. He provided more specific information to fishermen on the effect of size, sex and maturity on vitamin A content of liver by drawing attention to data published by the U.S. Fish and Wildlife Service.

FIG 21. Spiny dogfish being emptied from the cod-end on the deck of an American trawler (reproduced from *Pacific Fisherman*).

TABLE 11. Numbers of American vessels and boats licensed to use various types of gear with spiny dogfish-catching capability[a].

	Washington				Oregon			
Year	Set net	Set line	Otter trawl	Total	Set net	Set line	Otter trawl	Total
1937	—	419	43	462	—	60	4	64
38	—	577	61	538	—	120	1	121
39	—	544	70	614	—	106	7	113
1940	—	472	80	552	—	103	18	121
41	—	851	108	959	—	287	28	315
42	51	736	84	871	—	275	61	336
43	349	911	139	1399	127	185	65	377
44	503	850	236	1589	145	158	80	383
1945	403	826	255	1484	104	143	83	330
46[b]	—	—	—	(1162)	—	—	—	(321)
47	262	361	216	839	67	160	84	311
48	240	329	203	772	51	54	88	193
49	180	349	170	699	60	96	83	239
1950	44	289	164	497	17	46	65	128
51	34	228	151	413	5	70	59	184

[a]Source: Fisheries Statistics of the United States.
[b]Information not published. Figures in parentheses interpolated as mean of 1945 and 1947 values.

Bonham et al. (1949) in their excellent review of the dogfish fishery commented on a number of factors that determine the potency of dogfish liver: sex, size and maturity of the fish, color of the liver, and season of capture. Still later it was noted that even after the foregoing factors were taken into account, substantial variation was still to be found and Sanford et al. (1950) concluded that other factors as yet unidentified must be involved. The possibility that the age of a dogfish was a significant factor was mentioned early by Sanford and Bonham (1946) for they had observed a relation between liver color and the length of the second dorsal spine. Thirty years would elapse before the great variability in age for a given length was more fully appreciated.

e) Events of 1941–50

Having now reviewed a number of factors that influenced the course and conduct of the fishery it is appropriate to summarize events in the fishery itself. In 1941 Canadian production of dogfish (equivalent round weight) soared to a new record of nearly 14 000 t (Table 12), a large but unknown proportion (probably more than 7000 t) of which came from the Strait of Georgia. In Washington the landing was estimated at nearly 11 000 t for that year of which 6400 t were reported from "inside waters," i.e., Puget Sound, Straits of Georgia and Juan de Fuca (Table 12 and 13). Various indicators suggest that 1941 was the peak year of production from Canadian and U.S. inside waters and that total yield there probably was in excess of 12 000 t.

While there was little difficulty in arriving at approximate total removals by state or province, the actual fishing areas of origin became increasingly difficult to identify, particularly in respect to the Canadian fishery. Until 1939 the major areas of the Canadian catch could be fairly accurately identified with ports of landing. However, as fleets of vessels became increasingly mobile and wide-ranging halibut vessels and trawlers began landing livers in Vancouver from distant grounds, the usefulness of port-of-landing records deteriorated rapidly. Whereas in 1939, roughly 100% of the landings could be identified as to major area of catch, such allocation fell to 53% by 1941 and to as little as 27% in 1943–44 before steps were taken to improve statistical coverage of the fishing fleet. Even so, in only two succeeding years (1945 and 1946) did allocation to area of capture reach or exceed 50%. (For further details on the historical effectiveness of coverage the reader is referred to Appendix 6.) The same problem in varying degrees afflicted record keeping in the United States, the extreme example being California where except for isolated years it was impossible to separate dogfish landings from those of other sharks let alone identify areas of catch (Anon. 1949). Thus even the coastwide summary provided in Table 12 is not all together complete while Table 13, containing estimates of total

TABLE 12. Estimated total catch of spiny dogfish along the coast of North America from Southeast Alaska to Oregon (figures in tonnes)[a].

Year	Alaska	British Columbia	Washington	Oregon	Total	Year	Alaska	British Columbia	Washington	Oregon	Total
1935	—	3 484	126	?	3 610	1959	—	6 401	1 403	29	7 833
36	—	5 268	150	?	5 418	60	—	4 370	623	22	5 017
37	—	5 374	735	?	6 109	61	—	5 929	359	23	6 311
38	—	8 170	704	?	8 874	62	—	406	346	5	757
39	—	5 208	1 073	?	6 281	63	—	222	393	—	615
1940	7	8 790	1 517	564	10 878	64	—	982	833	—	1 815
41	290	13 965	10 877	2 633	27 765	1965	—	257	941	4	1 202
42	22	17 027	7 881	596	25 526	66	—	540	758	—	1 298
43	121	20 567	10 680	1 042	32 410	67	—	584	569	—	1 153
44	1 706	31 187	18 606	1 984	53 483	68	—	336	311	17	664
1945	503	23 373	10 621	1 007	35 504	69	—	1	272	25	298
46	548	11 417	10 039	1 410	23 414	1970	—	137	217	8	362
47	378	15 089	6 932	1 281	23 680	71	—	128	55	2	185
48	244	12 178	5 672	2 113	20 207	72	—	116	20	tr	138
49	503	16 010	4 829	1 553	22 895	73	—	5 056	6	tr	5 062
1950	8	2 213	875	319	3 415	74	—	1 070	749	11	1 830
51	5	4 000	1 112	69	5 186	1975	—	713	508	10	1 231
52	0	3 053	1 390	21	4 464	76	—	242	2 635	6	2 883
53	0	3 115	1 091	17	4 223	77	—	1 730	2 462	122	4 314
54	0	2 513	913	18	3 444	78	—	3 126	2 759	59	5 944
1955	0	2 621	878	—	3 499	79	—	4 757	4 284	344	9 385
56	0	1 124	692	24	1 840	1980	—	4 544	3 232	135	7 911
57	0	2 473	844	11	3 328	81	—	1 782[b]	2 185	?	3 967
58	0	1 606	1 920	29	3 555	82	—	3 914[b]	2 032	?	5 946

[a]Sources: Fisheries statistics of the United States (1935–76); U.S. statistics published by the Pacific Marine Fisheries Commission (1977–80); Wash. Dep. Fish. Stat. Rep. (1976–81). For Canada: Fisheries Statistics of Canada (1935–54); British Columbia catch statistics (1955–71); Smith (1972, 1981); Leaman (1982, 1983).

[b]Adjusted for unrecorded deliveries in USA (631 t in 1981; 1339 t in 1982).

Supplementary data. Preliminary figures for 1983: British Columbia, 3276 t; Washington, 1749 t; Oregon, unknown.

TABLE 13. Estimated landings equivalent round weight of spiny dogfish for American inshore waters of Puget Sound and vicinity (Area 4A) and Canadian waters of the Strait of Georgia and vicinity (Area 4B). Figures in tonnes[a].

	Statistical areas				Statistical areas		
Year	4A	4B	Total	Year	4A	4B	Total
1935	126	3 484	3 610	1960	577	3 432	4 009
36	150	5 268	5 418	61	386	2 234	2 620
37	735	5 135	5 870	62	346	406	752
38	704	7 243	7 947	63	393	222	615
39	1 073	4 498	5 571	64	833	982	1 815
1940	1 517	5 932+	7 449+	1965	941	248	1 189
41	6 428	6 138+	12 566+	66	758	531	1 289
42	2 372	4 554+	6 926+	67	569	584	1 153
43	3 567	3 582+	7 149+	68	311	336	647
44	3 796	1 102+	4 898+	69	272	1	273
1945	2 442	523+	2 965+	1970	217	133	350
46	2 299	658+	2 957+	71	55	128	183
47	1 989	747+	2 736+	72	19	106	125
48	2 484	389+	2 883+	73	6	4 295	4 301
49	1 371	510+	1 881+	74	749	1 035	1 784
1950	91	314+	405+	1975	491	680	1 180
51	288	484+	772+	76	2 583	239	2 822
52	414	638+	1 052+	77	2 366	1 637	4 003
1953	618	658+	1 276+	78	2 647	2 831	5 478
54	605	1 847	2 452	79	3 883	4 334	8 217
1955	536	1 705	2 241	1980	3 004	2 104[b]	5 108
56	518	416	934	81	(1 808)	1 212	3 020
57	51	1 115	1 166	82	1 944	2 011	3 955
58	1 768	987	2 755				
59	1 296	4 990	6 286				

[a]Source: For Canada: Fisheries statistics of Canada (1935–53); British Columbia catch statistics (1954–71). Smith (1973, 81); Leaman (1982, 1983); MS material 1979–82. For Washington: Wash. Dep. Fish. Stat. Rep. (1935–40; 1950–56, 1981); DiDonato (1974): 1941–49; PMFC statistics 1957–75; Pedersen and DiDonato (1982): 1976–80; Pattie and Tagart (1983): 1981; Wash. Dep. Fish. unpubl. data: 1982.

[b]Figures for 1981–82 adjusted for estimated unreported landings in USA (see Appendix 10 footnote).

Supplementary data. Preliminary figures for 1983: 1718 t in Area 4A and 2168 t in Area 4B.

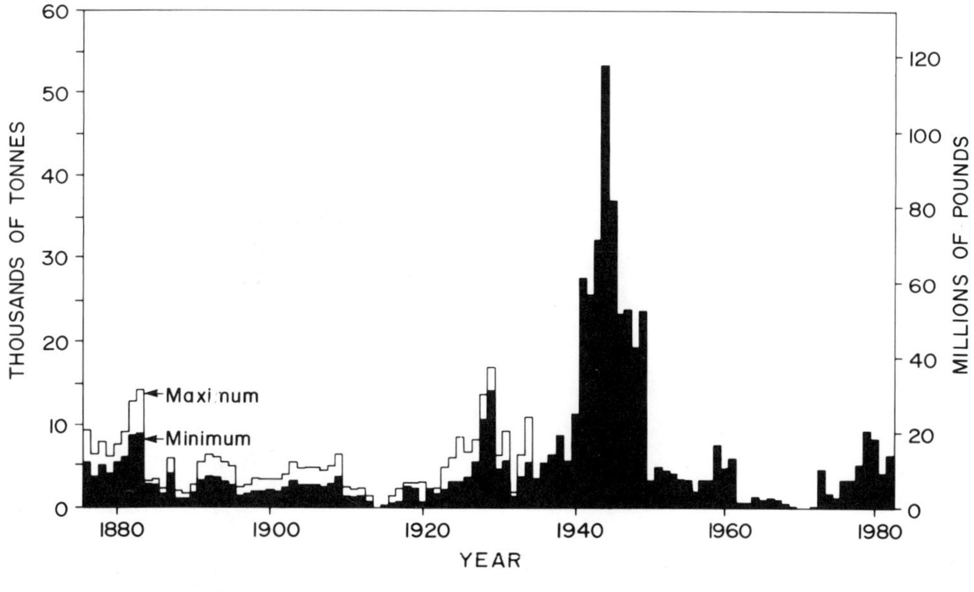

FIG 22. Long-term picture of production of spiny dogfish (equivalent round weight) in the Northeast Pacific from Alaska southward to Oregon.

production from Puget Sound and the Strait of Georgia, must be treated with caution in respect to the 1941–53 period.

In any event, the amount of spiny dogfish caught in the wide area from southeast Alaska to Oregon reached more than 53 000 t (118 000 000 lb) in 1944, a peak never approached before or since (Fig. 22) and one unlikely ever to be seen again. British Columbia accounted for 31 000 t or 58% of the total (Table 12). It was in that year that the lowly dogfish became the fourth most valuable species

TABLE 14. Landed value of the top 15 species of fish landed by Canadian fishermen in 1944[a,b].

Species	Region	Landed value
Cod	Atlantic	$8,366,000
Lobster	Atlantic	7,329,000
Herring	Atlantic	2,728,000
Grayfish (livers)	Pacific	2,662,000
Whitefish	Inland	2,607,000
Halibut	Pacific	2,232,000
Pink salmon	Pacific	2,046,000
Yellow pickerel	Inland	1,757,000
Sockeye salmon	Pacific	1,741,000
Chum salmon	Pacific	1,618,000
Mackerel	Atlantic	1,441,000
Herring	Pacific	1,392,000
Chinook salmon	Pacific	1,277,000
Haddock	Atlantic	1,138,000
Pilchard	Pacific	1,064,000

[a]Source: Fisheries Statistics of Canada published the Dominion Bureau of Statistics for species other than the Pacific salmon which had to be estimated by an indirect and approximate method.

[b]Excludes Newfoundland which had not yet joined the Canadian federation.

landed in all of Canada and the number one species in British Columbia. The value of livers alone was $2,662,000 (Table 14), which was exceeded only by three Atlantic species — cod, lobster, and herring. The high profile of the spiny dogfish in British Columbia in 1944 was partly attributable to the fact that it was a rather poor year for salmon and Pacific sardine production was down considerably from previous years. Still, the unusually great importance of dogfish relative to salmon prompted fishery offers to observe that

> ". . .particularly in the northern areas (of British Columbia) there was a lack of energy on the part of salmon gill-net fishermen to commence operations . . .(one factor being) the more attractive grayfish operations in the vicinity of the Queen Charlotte Islands and along the west coast of Vancouver Island" (Motherwell 1946).

Fishing grounds had indeed been expanded far from the relatively sheltered confines of the Strait of Georgia and Puget Sound. The Canadian gillnet fleet took a minimum of 6700 t from the shallow water of Hecate Strait while trawlers from both Canada and the USA removed substantial but unknown quantities from the deeper water. In 1944 many Washington vessels, including those from Oregon, concentrated off the British Columbia coast where high abundance offset reportedly lower unit potency of the liver oil. Friction developed as a result of U.S. vessels entering Canadian ports and some were charged with livering and discarding carcasses in inside waters (Pacific Fisherman Yearbook, 1945).

In October, 1944, fishery scientists and administrators from British Columbia and the Pacific coast states met in an informal conference at Victoria, B.C., to discuss the need for conservation of soupfin shark and dogfish. It was agreed that, because of a lack of suitable catch statistics, indisputable evidence of overfishing did not exist. Yet there were strong indications of trouble in the soupfin shark fishery and that dogfish catches had declined in such areas as the mouth of the Columbia River, Puget Sound, and the Strait of Georgia. In waters off the open coast of British Columbia it was believed

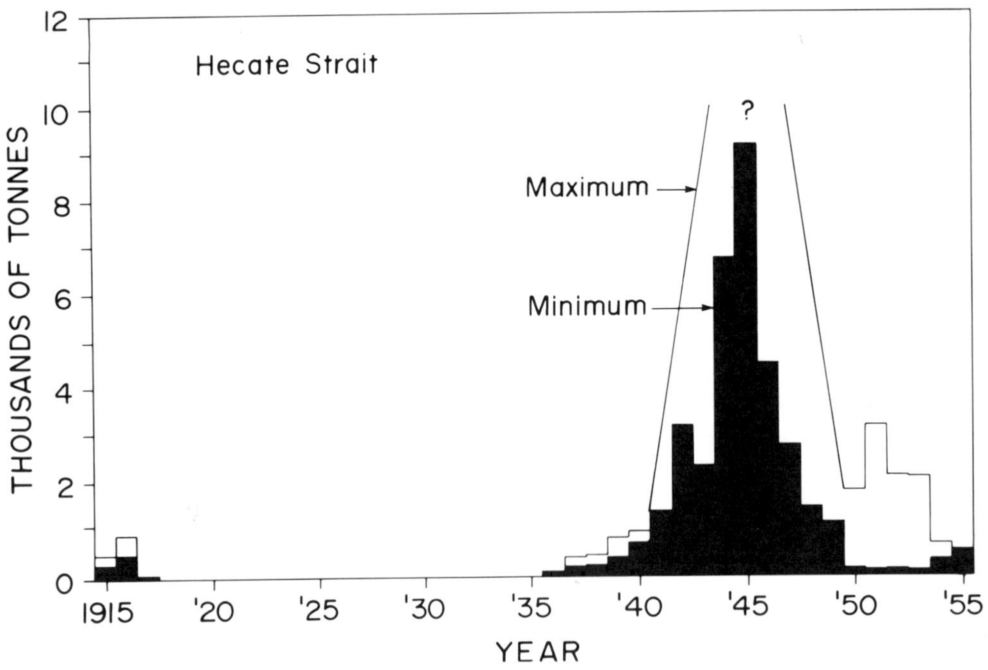

FIG 23. The 1915 to 1955 catch of spiny dogfish in Hecate Strait, British Columbia as estimated from liver and liver oil production of Canadian and American vessels.

TABLE 15. Known Canadian spiny dogfish production (equivalent round weight) by gear and by major areas of the British Columbia coast, 1943–48 (Figures in tonnes)[a,b].

	Strait of Georgia (Area 4B)			W. coast Vancouver Island (Areas 3C + 3D)				Queen Charlotte Snd. (Areas 5A + 5B)			Hecate Strait (Areas 5C + 5D)				Total known by gear and area
Year	Trawl	Line	All gear	Trawl	Gillnet	Line	All gear	Trawl	Line	All gear	Trawl	Gillnet	Line	All gear	
1943	—		3582		—	—	—	—	—	—	—	656	—	2136	5718
44	—		1102		556	—	556	—	—	—	—	6731	—	6731	8389
45	523	—	523	948	277	—	1225	—	—	—	3295	7017	—	10312	12060
46	585	73	658	1557	246	29	1832	247	310	557	769	3697	—	4466	7513
47	650	97	747	554	—	145	699	361	956	1317	1301	2294	35	3630	6393
48	288	111	399	493	—	127	620	240	746	986	1475	—	53	1528	3533
49	337	173	510	594	—	21	615	277	404	681	1065	—	25	1090	2896
50	254	60	314	177	—	1	178	1	121	122	57	—	—	57	671
51	433	51	484	277	—	9	286	35	446	481	75	—	—	75	1326
52	582	56	638	300	—	2	302	22	236	258	18	—	—	18	1216
53	634	24	658	426	—	9	435	—	223	223	38	—	—	38	1354

[a]Source: Fisheries statistics of Canada (1943–44); Barraclough 1948 (1943–47); and Pacific Biological Station records on the trawl fishery (1945–48).

[b]See also Appendix 6.

that the decline in availability of dogfish was negligible or small.

Rather little came of this meeting except for an agreement to exchange catch records and to publicize throughout the Canadian and U.S. fishing industries the desirability of avoiding wasteful practices in exploiting soupfin and dogfish stocks, specifically the capture of small fish whose livers were of relatively little value (Anon. 1944b). However, by 1944 no matter what other measures could have been adopted they would have been too little and too late.

Production in 1945 marked the start of a sharp decline, in part because the bonanza appeared to be over and ". . .the fishery was abandoned by many of the get-rich-quick operators" (Pacific Fisherman Yearbook, 1946). Also, stocks were declining and large dogfish were becoming increasingly difficult to find. Yet for Canadian fishermen 1945 was still a good year. From logbook records of the trawl fishery and a special study of the gillnet fishery in that year, it is known that a minimum of 10 300 t (22 million lb) of dogfish originated from Hecate Strait (Fig. 23 and Table 15) the last of the Pacific coast grounds to be exploited by North American fishermen.

f) Decline and Collapse

Although demand continued to be high in the late 1940s because of a shortage of supply and growing public acceptance of the merits of vitamin A (Pacific Fisherman Yearbook, 1947), total production in 1946 declined to less than half that in 1944, where it stayed for the succeeding three years (Table 12).

One of the few coastwide analyses to provide evidence of the rate of decline in stock size was that conducted by Barraclough (1948b) on the sunken gillnet fishery in Hecate Strait. His index of abundance, as shown in the following tabulation suggests that by 1945–47 the stock available to net gear had declined over 60%, while a progressive decrease in potency of liver oil signalled a decline in the average size of fish being caught.

Year	Abundance index	Average USP units of vitamin A per gram of oil
1943	1.000	?
1944	0.741	14,648
1945	0.367	13,183
1946	0.392	9,784
1947	0.376*	9,075*

*Unpublished data.

There was a brief drop in demand in mid-1947 on news that Japan had available and would be permitted to export a large amount of vitamin A oils (Pacific Fisherman Yearbook, 1948). Also, in the fall of that year an American company called Distillation Products announced that its scientists had succeeded in synthesizing vitamin A. This produced an instant drop in value of liver oil but it recovered on news that the synthesis had not yet reached a production stage.

The fishery continued moderately active until 1950 at which time the prospects of Japanese imports and commercial production of synthetic vitamin A became a realty and the market collapsed. Canadian production of dogfish liver, now incidental to foodfish fisheries for other species (halibut, Pacific cod, lingcod, and several flounder species) drifted from 1950 to 1958 at low levels between 130 and 450 t annually. Price had fallen to 10–12¢/lb where it had been in 1941 at the start of the dogfish bonanza.

To gain some appreciation of the size of the spiny dogfish resource (medium to large sized individuals) in the northeast Pacific at the start of 1941, let us make the not unreasonable assumption that the increase in weight of the stocks through growth and recruitment (both very slow processes) over, say, 9 years was balanced by the likewise slow losses due to natural death. A summation

TABLE 16. Estimates of the amount of spiny dogfish caught and discarded during Canadian trawling operations for groundfish, based on interviews with vessel captains in 1958[a].

Fishing area	Number of trips	Tonnes caught and discarded			
		Total	Average per trip	Minimum	Maximum
Strait of Georgia	7	70	10.0	0.5	18.0
Lower west coast of Vancouver Island	47	499	10.6	0.0	101.0
Upper west coast of Vancouver Island	5	22	5.5	0.2	15.0
Cape Scott	12	3	0.2	0.0	2.0
Goose Island	6	1	0.2	0.0	0.7

[a]Source: Pacific Biological Station interview records.

of catches from 1941 to 1949 (see Table 12), shows that the weight of the stock at the beginning of 1941, had to be no less than 265 000 tonnes or 584 million lb. It is a minimum estimate because by no means all of the fish present at the beginning of 1941 had been caught by the end of 1949 and we are unable, through lack of information, to include catches that were made off California. Actually, a more refined approach to estimation of initial stock size (using the method of Leslie and Davis 1939) provides figures ranging from 240 000 to 390 000 t, and these applied to the open coast of North America alone and only to the marketable stock (Ketchen 1969; Ketchen, unpublished data). If we were to make allowance for the stock(s) in inshore waters, the total for the Northeast Pacific as a whole probably lay in the range of 300 000–500 000 t.

5. The Government-Assisted Fishing Era (1951-74)

As already noted, once the fishery for spiny dogfish collapsed in 1950, landings of dogfish liver continued to be made for a number of years but at much lower levels than formerly. Coastwide round-weight equivalences of dogfish landed ranged downward from about 5200 t in 1951 to little more than 1800 t in 1956 (Table 12). Half to three-quarters of the production came from British Columbia waters, mainly as an incidental product of the trawl fishery for other species of groundfish.

a) The Pestilence Returns

It was not until about 1958 that the Fisheries Research Board of Canada (FRB), through its port representatives began to receive complaints about the growing nuisance of dogfish on many of the important trawling grounds. A special survey of trawler skippers operating off the southern half of the British Columbia coast showed the lower west coast of Vancouver Island (Area 3C) and the Strait of Georgia (Area 4B) to be the areas where interference and wear-and-tear on gear were greatest (Table 16). Although the average discard was only 10 t per trip, these were clearly underestimates of the quantities of dogfish on the grounds, for fishermen were avoiding grounds, depths and time of day where and when concentrations were so great as to preclude fishing entirely. Shepard and Stevenson (1956) ventured a guess that the magnitude of the rebuilding spiny dogfish stocks was somewhere between 225 000 and 450 000 tons.

There were also increasing complaints about damage and destruction of gear from fishermen using set-lines, herring and salmon purse-seines, and salmon gill nets. Added to this problem was the widely held conviction that predation by dogfish was of such intensity as to reduce significantly the abundance of other, more valuable species. In due course representations reached Canadian Department of Fisheries headquarters in the form of a request for a subsidy on dogfish liver, to encourage resumption of an economical fishery and consequent reduction in the number of dogfish.

At about the same time and seemingly independent of events in Canada, there was public clamor for some sort of action in the State of Washington to rid fishing grounds of the dogfish menace (Clark 1958; Anon. 1958; Wedin and Moore 1959). On September 2, 1958, the 85th Congress of the United States passed Public Law 85-887 (S-2719) "authorizing and directing the Secretary of the Interior to investigate and eradicate the predatory dogfish sharks to control the depredations of this species on the fisheries of the United States and for other purposes". A three-man committee representing the fishing industry and the state and federal governments was struck to develop a program of action which included instructions to conduct ". . . (1) a review of available information with supporting data concerning effects of the dogfish populations on the commercial fisheries, abundance and distribution of Pacific dogfish, and past experience in utilizing and controlling this species," and (2) (to make) ". . .recommendations for carrying out a dogfish research and control program."

Part of the committee's subsequent report included a review of the biology of dogfish, the history of the fishery, methods of control, technological and other considerations (Alverson, 1958). A revised and more formal edition appeared at a later date (Alverson and Stansby 1963). The committee's report also included estimates that the annual cost to fishermen of damage inflicted on their gear by incidental catches of dogfish was in the vicinity of $2 million (salmon gillnet fishery — $600,000; troll — $500,000; purse-seine — $68,000; halibut set-line fishery — $450,000; groundfish trawl fishery — $250,000; sports fishery — $350,000).

Recommendations of the committee included, in part, requests for (1) $250,000 for control of dogfish along the Pacific coast (in cooperation with Canada), (2) $95,000 for study of ways and means to provide for control, utilization and marketing of dogfish, (3) provision in a control program for making "incentive payments" not exceeding \$15 per ton of round dogfish or 15¢/lb of liver, (4) exploration of domestic and foreign market possibilities, and (5) imposition of increased rates of duty on imports of foreign fish oils and by-products (for more complete details see Anon. 1958).

As evident from United States fisheries statistics (Tables 12 and 13) funds requested in support of these recommendations were not forthcoming. Although dogfish utilization rose to 1920 and 1400 t in 1958 and 1959, respectively, from 700 to 900 t in years immediately preceding, it fell to new low levels in subsequent years.

b) Canadian Dogfish Control/Utilization Programs

In contrast to events across the border, the request for federal government assistance met with a favorable response from Canada's Minister of Fisheries, the Honorable J. A. MacLean. Between 1959 and 1962, funds were provided for several attempts to introduce control measures and/or encourage resumption of commercial fishing for dogfish.

i) 1959 Project — Early in January 1959, the Minister allocated \$250,000 for the first project which had two components: (1) four trawlers were to be chartered simply to kill dogfish during January 19–March 29 and paid a certain amount per ton with a bonus for every ton over 300 t caught in the Strait of Georgia, and (2) other vessels (mainly trawlers and set-liners) would be paid 10¢/lb for liver collected in the Strait of Georgia *and* any other area of the coast, within the period January 9 to March 31, 1959. An additional subsidy of 5¢/lb on liver was to be paid to fishing companies participating in the program.

On July 24 of the same year, the Minister allocated an additional \$250,000 and this was to be applied as a 10¢ subsidy on liver only, and with no special provision for the Strait of Georgia during July 24, 1959 and March 31, 1960.

Details of the January–March operation in the Strait of Georgia alone may be found in Appendix 8, but summary data by calendar year are provided in Table 17 along with information for other years. This shows that the total landing (equivalent round wt) in 1959 was 6401 t of which 4990 t or 78% originated from the Strait of Georgia. The latter figure includes the charter boat or "killer" operation which accounted for 728 t. The dominant fishing gear was, as in pre-war years, the baited set-line, followed by trawl and in turn by "other gear" (hand-lines and floating gill nets).

ii) 1960 Project — At the beginning of January, 1960 there were approximately \$58,000 remaining of the funds allotted in the previous calendar year. Thus the liver subsidy remained in effect until the end of the fiscal year (March 31, 1960) by which time 255 t of liver (2260 t round wt) had been landed.

On October 11, 1960 the Minister announced that further support would be given to the fishery and allotted \$150,000, from which fishermen were to be paid a subsidy of 12¢/lb of liver. From that date until the end of the calendar year an additional 239 t of liver were landed, leaving a carry-over of \$85,000 for the remaining months of the fiscal year. Thus the total landing of dogfish (equivalent round wt) for 1960 was 4370 t (Table 17) of which 79% or 3432 t originated from the Strait of Georgia.

iii) 1961–62 Project — As in 1960 the subsidy funds in calendar year 1961 consisted of a carry-over of funds from the 1960–61 fiscal year plus an additional \$150,000 allotted by the Minister of Fisheries on April 17, 1961

TABLE 17. Estimation of the equivalent round weight (tonnes) of spiny dogfish caught in major areas of the British Columbia coast by various types of gear during 1954–63[a].

Fishing area	Stat. area	Gear	1954	1955	1956	1957	1958	1959	1960	1961	1962	1963
Hecate Strait	5C + 5D	Longline	214.4	215.0	245.7	149.8	93.5	223.5	—	—	—	—
		Trawl	24.7	150.7	308.1	519.6	231.2	428.0	—	—	—	—
		Other	58.8	161.2								
		Total	297.9	526.9	621.5	728.0	368.3	682.1	782.8	559.2	—	—
Queen Charlotte Sound	5A + 5B	Longline	—	68.4	13.3	—	—	57.2	—	—	—	—
		Trawl	7.7	11.9	11.8	76.2	62.5	29.0	—	—	—	—
		Other	0.4	—	—	—	—	2.4	—	—	—	—
		Total	8.1	80.3	25.1	76.2	62.5	88.6	82.7	65.0	—	—
Strait of Georgia	4B	Longline	178.6	157.8	5.2	132.0	68.6	2731.0	—	—	—	—
		Trawl	917.3	1047.7	330.8	798.1	819.4	1806.7	—	—	—	—
		Other	751.5	499.8	79.8	183.8	99.0	452.2	—	—	—	—
		Total	1847.4	1705.3	415.8	1113.9	987.0	4990.0	3432.2	2234.0	—	—
W. Coast Vancouver Island	3C + 3D	Longline	10.6	17.8	0.3	—	0.4	1.6	—	—	—	—
		Trawl	344.4	277.0	59.2	552.7	186.1	628.2	—	—	—	—
		Other	4.6	14.1	2.0	2.0	1.3	10.9	—	—	—	—
		Total	359.6	308.9	61.5	554.7	187.8	640.7	72.3	3071.0	—	—
Total		Longline	403.6	459.0	264.5	281.8	162.5	3013.3	—	—	—	—
		Trawl	1294.1	1487.3	709.9	1946.6	1299.2	2891.9	—	—	—	—
		Other	815.3	675.1	149.5	244.4	143.9	496.1	—	—	—	—
		Total	2513.0	2621.4	1123.9	2472.8	1605.6	6401.3	4370.0	5929.2	406.0	221.8

[a]Source: Unpublished records of the Department of Fisheries, Vancouver, B.C.

for the 1961–62 fiscal years. The total landing (equivalent round wt) was 5929 t, but in contrast with the previous year, only 38% of this was taken from the Strait of Georgia. Fishermen reportedly were encountering difficulty finding spiny dogfish in sufficient numbers in the Strait and were turning to the west coast of Vancouver Island as an alternative source of supply (Table 17).

The subsidy program in calendar year 1962 was nothing more than an extension of decisions made for the 1961–62 fiscal year. As of January 1, 1962 little more than $12,000 remained from the original allocation of $150,000. By the end of March, a landing of 46 t of liver, equivalent to 406 t round weight and all from the Strait of Georgia used up the remaining subsidy fund, thus ending the program.

iv) Assessment of the 1959–62 programs — There is no question that introduction of the federal government's incentive programs during 1959–62 did much to encourage exploitation of dogfish, particularly in the Strait of Georgia. Removals from that area in 1959–61 far exceeded those of the preceding five years and probably in all of the years following the collapse of the fishery in 1950. As a "control" measure the programs lost some of their impact by the political necessity to make the subsidy on livers coastwide in its application (i.e., to give equal opportunity to fishermen living in Prince Rupert — too far away to be involved in a Strait of Georgia fishery). More serious however, was the move by south coast trawl fishermen in 1961 to seek dogfish belonging to the offshore stock(s) along the west coast of Vancouver Island. To achieve a "control" objective in respect to offshore stock(s) was a practical impossibility without the active cooperation of the United States. This shift in emphasis was a major factor in the administrative decision to terminate the subsidy payments at the end of the 1961–62 fiscal year.

One would have expected that the removal of 10 700 t (more than 24 million lb) by the end of 1961 would have reduced measurably the size of the spiny dogfish stock in the Canadian part of the Strait of Georgia. While this may well have been the case, it was not evident from available statistics on the fishery. Examintion of records of catch and fishing effort for the trawl fishery (Table 18), shows that whereas the catch rate (CPUE) in 1960 was only about half as good as in 1959, it was better in 1961 than that in 1959 largely because of the exceptionally high fishing success in the October–December period. This points up one of the frequently encountered difficulties of using catch and effort data to measure changes in relative abundance, namely, that the underlying assumption of constant availability from year to year is not always satisfied. Availability (technically defined as the fraction of a fish population which lives in regions where it is susceptible to fishing during a given fishing season) can be expected to vary from season to season, particularly for any species that tends to move about in schools of varying density. It is apparent in Table 18 that there are seasonal differences in dogfish availability, perhaps related to the breeding cycle, for CPUE in all 3 years was higher in October–December than in January–March. In 1961 spiny dogfish were obviously much more vulnerable to capture in October–December than in the same months of preceding years, thus destroying any basis for quantifying the impact of the subsidized fisheries in 1959–61 on the Strait of Georgia stock.

TABLE 18. Statistics of spiny dogfish catch (t), fishing effort (h) and catch per unit of effort (kg per h), by Canadian trawlers (mainly 25–49 GT class) operating in the Strait of Georgia[a].

		Jan.-Mar.	Oct.-Dec.	Total
1959	Catch	286	211	497
	Effort	3365	1400	4765
	CPUE	85	151	104
1960	Catch	94	80	174
	Effort	2765	1046	3811
	CPUE	34	76	46
1961	Catch	157	221	378
	Effort	2509	696	3105
	CPUE	63	317	122

[a]These records apply only to vessels whose captains were interviewed or who provided logbook records of their operations.

c) Government Assistance in Developing a Foodfish Fishery for Dogfish in British Columbia

i) First steps — Following the subsidized liver fishery that ended in March 1962, there were no further Canadian government interventions until January 1966. At that time the Minister of Fisheries set aside $25,000 for an experimental dogfish marketing program. Contracts were drawn up with two fishing companies, with the provision that the Department of Fisheries would cover losses, while at the same time arrangements were made with a number of set-line and trawl fishermen to land round dogfish at $60 and $40 per ton, respectively. A survey of world markets revealed an interest in skinned belly-flaps by companies in West Germany and Belgium, in whole fish for the fish-and-chips trade in Great Britain and whole fish for the production of "kameboko" (a minced fishcake) in Japan. The products were well received except in Great Britain where one wholesaler ruled out sales on grounds that the frozen flesh lacked the pink color to be found in the European (Norwegian) imported product. Only $11,000 of the allocation was used during the January–March period and only 160 t of dogfish were used in the experiment. However, this was sufficient to demonstrate some economic prospect providing the high costs of processing and shipping could be reduced.

Although the cost factor would continue later to inhibit development of markets in Japan, it is of historical interest to digress briefly with a comment on the shortage of supply of spiny dogfish in Japanese home waters. Immediately after World War II Japanese fishermen began intensive fishing for production of livers and vitamin A — the same product that was introduced on the North American market and contributed to the collapse of the fishery in the Northeast Pacific in 1950 (p. 43). By 1952 Japanese landings (round weight) reached a peak of 60 000 t and then declined rapidly to 30 000 t by 1957 and to even lower levels (ca. 12 000 t) by 1964. This was essentially a replication both in timing and magnitude of events a decade earlier in North American waters and provided another example of the difficulty of developing sustainable fisheries on a species with low growth and reproductive rates. Out of the initial Japanese liver-oriented fishery grew the fishery for dogfish as a foodfish, i.e., as one of the sources of supply for the new and expanding demand for kameboko. In 1964 and in Tokyo alone there were 870 kameboko factories producing close to 260 000 t of this product annually (Y. Kitano, Japan Fisheries Agency, personal communication). Although spiny dogfish was a minor contributor, many producers preferred it over other species. Thus during the 1966–69 period there were numerous private and intergovernmental enquiries into the availability of dogfish and cost of supplying the Japanese market from unutilized stocks in Canadian waters.

Returning to the Canadian assistance program, the Department embarked on the second phase of the experiment in November, 1966 with the announcement that fishing companies would be paid 11¢/lb for all skinned belly-flaps produced up to a maximum of 220 000 lb (100 t). This assistance was to be paid on the understanding that the companies would pay fishermen $50 per t of dogfish landed in the round or 13.5¢/lb for unskinned flaps. Between November 1966 and March 1967, 667 t taken in the Strait of Georgia were used for the production of flaps and a small quantity of skinned bodies. Sales were made in West Germany and Great Britain.

However, from the trend in Canadian landings between 1966 and 1972 (Tables 12 and 13) it is apparent that nothing of substance developed from the marketing experiment. In 1969, the total landing of dogfish in British Columbia and Washington fell to 273 t and to even lower levels in 1971–72 — lower than anything reported since 1915.

ii) The mercury problem — It was in the late 1960s and early 1970s that a new problem arose which, at least temporarily, made dogfish even less economically attractive than usual, namely, the discovery that the flesh of that species contained high concentrations of mercury. It is important to recall the effect this contaminant had on markets for numerous species of fish in North America.

In the 1960s world attention was drawn to Minamata Bay, Japan, where mercury-containing effluent from an industrial plant contaminated marine fishes and invertebrates used as food by local inhabitants. The severe symptoms that developed from this poisoning became known as "Minamata disease".

In North America, public awareness of the extent of environmental mercury contamination developed almost explosively as the result of a Canadian announcement on March 24, 1970 that commercially caught fish from Lake St. Clair, Ontario, were unfit for sale. The discovery followed upon earlier (1969) revelations of the presence of increased mercury levels in wildfowl in Alberta which had eaten mercury-treated grain, and in fish of the Saskatchewan River (Mastromatteo and Sutherland 1972). In both Canada and the United States this set off widespread investigation of the commercial and recreational fishes of the Great Lakes Basin and a search for sources of contamination. Highest concentrations were found in the vicinity of industrial cities and plants, and among the larger, fish-eating species. The reason for the latter was that mercury, once ingested, is so slow to be excreted that longer-lived species whose diet consists of contaminated smaller fishes or organisms become accumulators of the heavy metal. Moreover, mercury is one of the few metals that tend to accumulate in the muscle of fish.

In Canada, the federal Food and Drug Directorate and in the U.S., the Food and Drug Administration adopted a concentration of 0.5 parts per million (ppm), wet weight, in fish muscle as a tentative maximum for commercial fishes. All fishes exceeding this level were to be destroyed. In both countries there ensued numerous fishery closures and bans on marketing of certain species or certain species over a specified maximum size.

The testing of fish products for mercury extended to marine regions where it was found that, even among large individuals of far-ranging oceanic species such as swordfish and tunas, and among large halibut on coastal banks far from industrialized areas, mercury was present in amounts exceeding the 0.5 ppm limit. This revelation together with the discovery that freshwater fishes in areas remote from civilization were similarly afflicted, showed that mercury occurs in the natural environment as well as in areas subjected to industrial pollution.

Not surprisingly, the slow-growing, long-lived spiny dogfish of the Northeast Pacific proved to be a great accumulator of mercury. For the Strait of Georgia, concentrations far in excess of 0.5 (in some cases as high as 2.0 ppm) were found among the majority of males over 70 cm in length and females over 90 cm in length (Forrester et al. 1972). It was later determined that all female fish over estimated ages of 32 yr registered loads greater than 0.5 ppm (Ketchen 1975).

Similarly high levels were reported by Hall et al. (1977) in samples of fillets and belly-flaps of dogfish collected in various localities of Puget Sound. Both Canadian and American teams of investigators detected differences among localities and suggested these were due to variations in the level of local industrial pollution. Hall et al. (1977), compared their results with those of Childs et al. (1972) for the Oregon coast, which showed relatively low levels of mercury. They took this as corroboration for the previously referred to conclusion (p. 7) that stocks of spiny dogfish in inshore waters are relatively separate from those off the open coast.

In Canada, the revelation of high levels of mercury in dogfish came in January 1971 at a time when the fishery in the Strait of Georgia was at its lowest ebb (Table 13) and at a time when a private-sector attempt was being made to export dogfish flesh for human consumption. This slowed the development of a significant foodfish fishery. Incidentally it was also to delay or complicate assessment of the economic feasibility of sablefish culture, for Kennedy (1972) had demonstrated the merits of dogfish flesh as a food for pond-reared juveniles.

In due course the Food and Drug Directorate in Canada and the Food and Drug Administration in the USA were to soften their positions in respect to fish flesh carrying a mercury load of more than 0.5 and 1.0 ppm, respectively. They recommend that consumers exercise moderation in eating species prone to high mercury levels. In regard to exports, the Canadian Fish Inspection Branch (Pacific Region) now makes periodic tests to determine the average mercury level in products to be shipped abroad. Standards regarding the maximum permissible mercury content vary from country to country. The most tolerant countries (1.0 ppm) are Denmark, Federal Republic of Germany, Finland, Sweden, United Kingdom, and USA; and the least tolerant (0.5 ppm) are

Belgium Netherlands, Norway, Poland, and Portugal (G. Grigg, Fish Inspection Branch, Vancouver, personal communication). In all cases the tolerance level is determined by the average in samples rather than by the maximum in such samples.

iii) The 1973 subsidy program — On January 25, 1973 another attempt was made by the Canadian Department of Fisheries to reduce the numbers of dogfish and at the same time create a viable processing industry. The program as announced by the Minister, the Honorable Jack Davis, provided $250,000 for this purpose and contained unique terms of reference. Two tons of roe herring (a new and lucrative fishery development) would be earned for each ton of dogfish purchased by a fishing company. Participating companies would receive an added incentive of $50 per ton of dogfish towards the cost of processing. The subsidy was earmarked primarily for the Strait of Georgia with only $50,000 of the total amount being assigned to dogfish catches in north coast waters. The program was to run from date of announcement to March 31, 1973 or until funds were exhausted, whichever came first.

Results of the program are summarized below from a more comprehensive document prepared by McEachern and Roberts (1973):

Landings. During February–March, 1973, 3560 t of spiny dogfish were removed from the Strait of Georgia — 66% by set-line vessels and 28% by trawlers, while the remaining 214 t were taken mainly by hand-line vessels but included 65 t caught with sunken gill nets under special permit. In the Strait of Georgia the set-line catch was made primarily in Minor Statistical Areas (MSA) 13, 16, and 17 and the trawl catch in MSAs 14 and 17 (Fig. 3), i.e., more or less in the traditional pattern of fishing areas occupied during preceding decades.

Off the north coast the catch was made almost entirely on the eastern side of Hecate Strait near Browning Entrance (Major Area 5D, Fig. 2) and amounted to 689 t of which 490 t were taken by trawlers.

Processing and marketing. McEachern and Roberts (1973) reported that of the 4249 t of dogfish landed from all areas, the "finished" product amounted to 30% or 1268 t. The amount of each product is given in the following tabulation:

Product	000s kg	Product	000s kg
Frozen round	151.8	Liver	7.4
Frozen — dressed	274.9	Fishmeal	409.3
— head off	147.0	Oil (liver & body)	225.1
Belly-flaps	21.2	Minkfood or	
Body meat	8.5	fertilizer	22.7
Body fillets	0.2	Total	1268.1

Markets for whole and dressed spiny dogfish were found in Japan only because of an agreement with Canadian companies which made purchase of herring roe conditional upon the purchase of a certain amount of dogfish. Looking to the future, McEachern and Roberts (1973) concluded that the prospect of selling dogfish to Japan even in a dressed-head-off state was very limited. Besides, that form of product would not generate much shore employment in Canada. Processing of dogfish as fishmeal, despite the high prices being paid on the world market at that time, showed little possibility of ever being economic in its own right.

What did seem to be encouraging was the prospect of gaining entry to the European market for belly-flaps and body meat. The latter was used in the "fish and chips" trade in Britain; belly-flaps and some quantities of body meat were in demand in West Germany, while body meat alone had prospective markets in Mediterranean countries. There also appeared to be some opportunity to supply the limited but high-priced markets of the Orient with dogfish fins and tails for use in the "sharkfin soup" trade. It was apparent, however, that a viable operation would depend on improvement in processing and storage methods and on restriction of the trade to dogfish of relatively large size.

Many difficulties encountered in 1973 in the processing and marketing sectors could be blamed on the short notice given to the subsidy program and the shortness of the period of operation itself which gave processors little or no lead time to prepare for the new methods required for producing belly-flaps and body meat. "Most processors did not have the facilities to handle both dogfish and herring roe simultaneously so were forced to give the dogfish away or incur the high costs of freezing for use after the herring fishery had ended. Under these circumstances it could be expected that market experimentation would not be successful." (McEachern and Roberts 1973).

A high percentage of the dogfish caught was reduced to fishmeal which gave low return for money expended. However, companies that bought dogfish for this purpose did so in order to obtain earned herring roe quotas, not to develop new products and find markets.

Nevertheless the 1973 program appeared to be effective from the standpoint of creating employment. Also on the plus side was the experience gained by processors in handling and preparing dogfish products, and in seeking suitable markets. During the remainder of 1973, following termination of the subsidy program, an additional 807 t of dogfish were landed in British Columbia, bringing the total for the year to 5056 t of which 4295 t (85%) came from the Strait of Georgia (Tables 13 and 19).

iv) The 1974 subsidy program — Federal government assistance was resumed at the beginning of fiscal year 1974–75 but at a lower level. A total of $50,000 was set aside, again for a subsidy of $50 per ton payable to processors, for a total landing of 1000 tons. Eighty-five percent of this limit was achieved by the end of the calendar year. All but a small fraction of the catch was made in the Strait of Georgia and within that area most of the activity involved set-line vessels in the northern section of the Gulf Islands (minor statistical area 17).

In 1975 and 1976, Canadian production, though still higher than in the previous decade, declined sharply for want of sufficient incentive. However, some private ventures to enter the European market continued at a low level.

The requirement of relatively large fish for foodfish processing prompted some Canadian fishermen to seek legalization of sunken gill nets in the Strait of Georgia. Between November 1975 and March 1976, a few special

TABLE 19. Canadian landings (equivalent round weight) of spiny dogfish by major statistical areas of the British Columbia coast, 1964–75 (figure in tonnes).

	Area 4B	Areas 3C + 3D	Areas 5A + 5B	Areas 5C + 5D	
Year	Strait of Georgia	West Coast Vancouver Island	Queen Charlotte Sound	Hecate Strait	All areas total
1964	982	—	—	—	982
65	248	—	9	—	257
66	531	tr	9	—	540
67	584	—	—	—	584
68	336	—	—	—	336
69	1	—	—	—	1
1970	133	4	—	—	137
71	128	—	—	—	128
72	106	10	—	—	116
73	4295	17	2	742	5056
74	1035	35	—	tr	1070
1975	680	32	—	—	713

permits were issued by Fisheries Management and although total production during that period was only 1331 t, 24% was taken with gill nets. However, the method proved laborious for those fishermen involved because the gear was made of discarded salmon gill nets which were not particularly suited for bottom fishing. While there was some interest in constructing gear specifically for dogfish, no permits were issued after March 1976 because of apprehension on the part of Fisheries Management that the incidental capture of salmon and lingcod would be unduly high. No up-to-date information was available on this delicate subject but a limited study by Barraclough (1948a) had demonstrated that lingcod were particularly vulnerable to capture when nets were set on reefs.

d) Government Assistance in the State of Washington

As already mentioned, an attempt to obtain government assistance in eradicating the dogfish nuisance in 1958 proved unsuccessful (p. 45). Production fluctuated at a low level through the 1960s and by 1973 had fallen to virtually nil (Tables 12 and 13). Late in that year, State government assistance came in the form of a $125,000 allocation to the Washington Department of Fisheries to test the feasibility of using dogfish meal in the State's salmon and trout hatchery diets. This action came not in response to a request from fishermen looking for relief from the dogfish nuisance but from concern over the increasing cost of fishmeal and potential shortages of herring and other traditional fishmeal sources in the future. Between January and mid-April, 1974 fishermen were paid $50 per ton of dogfish to supplement their regular income from foodfish species, and this was raised to $70 for the remainder of the fishery through early June. Landings by 22 trawlers and three gillnet vessels operating almost exclusively in North Sound (MSA 80 in Fig. 3), amounted to 232 t which in turn yielded 32 t of fishmeal (DiDonato and Westgard 1974).

Feeding experiments were subsequently conducted with dogfish meal as a component of the so-called Oregon Moist Pellet (OMP) and the results proved satisfactory, although there was some concern about the mercury content and its possible deleterious effects on young salmon. The level ranged from 0.71 to 2.05 ppm in the meal but considerably lower (0.14–0.24 ppm) in the OMP (DiDonato and Westgard 1974). Even so the proportion of dogfish meal in OMP is today limited by the mercury level which seems safe for salmon fry.

Apparently meal production continued after the period of incentive payments because the total dogfish landing for the year, almost all from Puget Sound, amounted to 749 t (Anon. 1981). In 1975 dogfish destined for reduction and landed by trawlers alone was 283 t and in succeeding years through 1981 ranged from three to 124 t (Pattie and Tagart 1983).

6. Coming of the Foodfish Era (1975 to present)

Traditionally, North Americans have avoided using spiny dogfish as a foodfish more for aesthetic reasons than because of problems of processing or palatability. In Europe, on the other hand, the species has long been regarded as a relatively inexpensive source of protein in a variety of forms, and until the last decade the entire catch by Norway and the United Kingdom found a market in the fish-and-chips trade in London and southeastern England (Holden 1977). Demand for the "backs" has now spread to Belgium, France, and other countries, while "belly-flaps" are smoked, cut into thin strips and served as "see-aal" (sea eel) in West Germany. The various stages of processing and packaging are shown in Fig. 24.

Traditional sources of supply have been from waters of northwestern Europe, and particularly from what Holden describes as the Scottish–Norwegian stock. The areas occupied by this stock lie to the north and west of Scotland, along the coast of Norway and in the intervening waters of the northern North Sea. Production from these areas reached a peak of about 42 000 t in 1963 and entered a long decline through 1980 when the total landing was less than a third of that in 1964 (Fig. 25). The decline was accurately predicted by Holden and Meadows (1964) on the strength of evidence that at the then prevailing rate of exploitation the Scottish–Norwegian stock was unable to replace itself because of the species' low fecundity and reproduction rate.

The decline in landings to less than 14 000 t annually by 1980, is presumed to be one of the reasons, if not the main one, that the door has opened to imports from both the east and west coasts of North America.

In spite of the 1973 government-subsidized experiment in processing spiny dogfish as a foodfish in British Columbia and the supposed "head start" that this provided, it was the State of Washington fishing industry that first grasped the opportunity to supply the British and European markets. As subsequent developments in Washington and British Columbia followed different courses, it is best that they be treated separately.

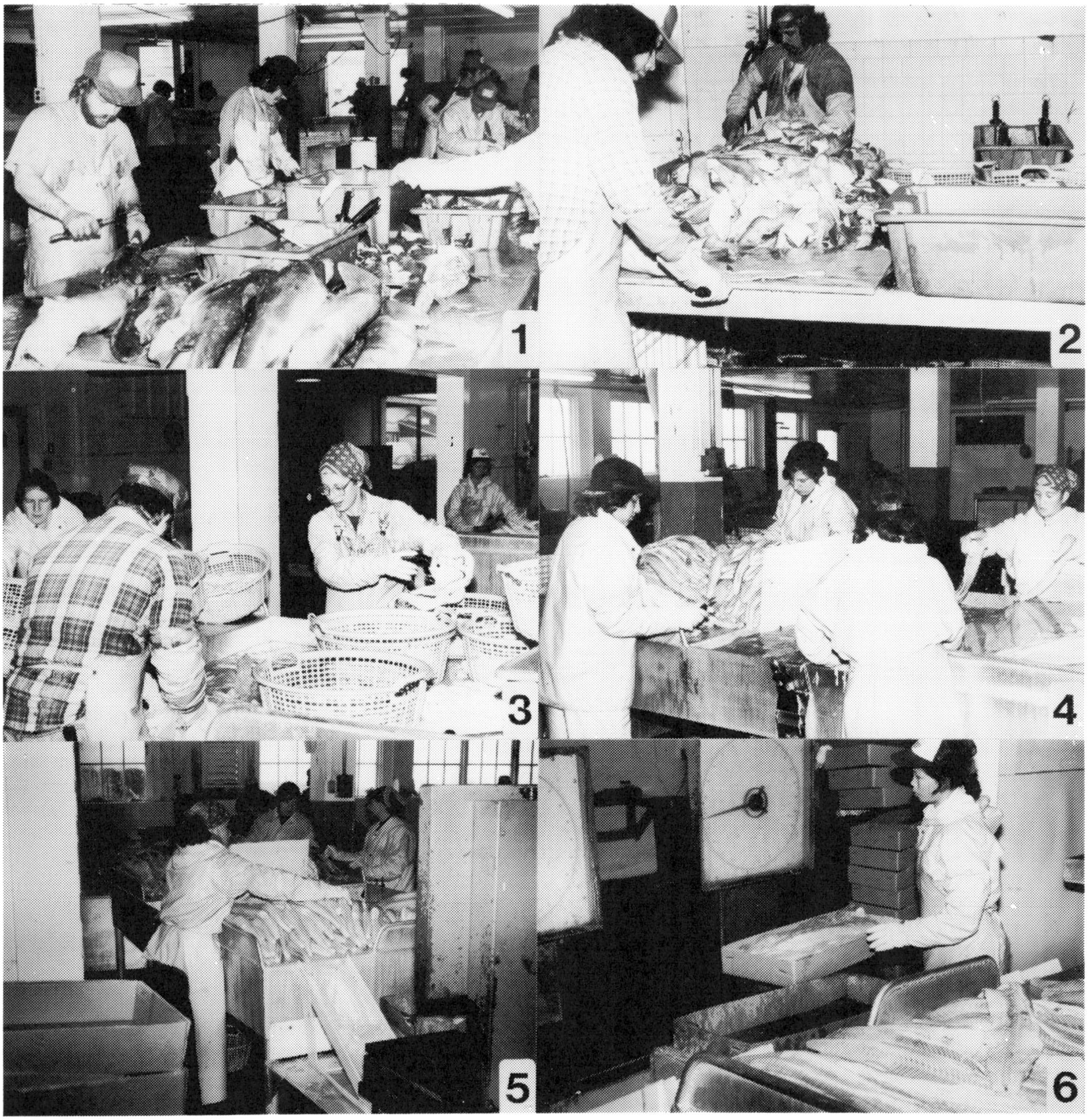

FIG 24. Stages in the processing of spiny dogfish for export of foodfish products to Europe: (1) gutting and skinning operation; (2) filleting of belly-flaps, (3) washing of belly-flaps, (4) washing of backs, (5) individual packaging of backs and (6) boxing of backs for freezing and shipping. Photographs by author, through courtesy of Arrowac Fisheries, Bellinghan, Washington.

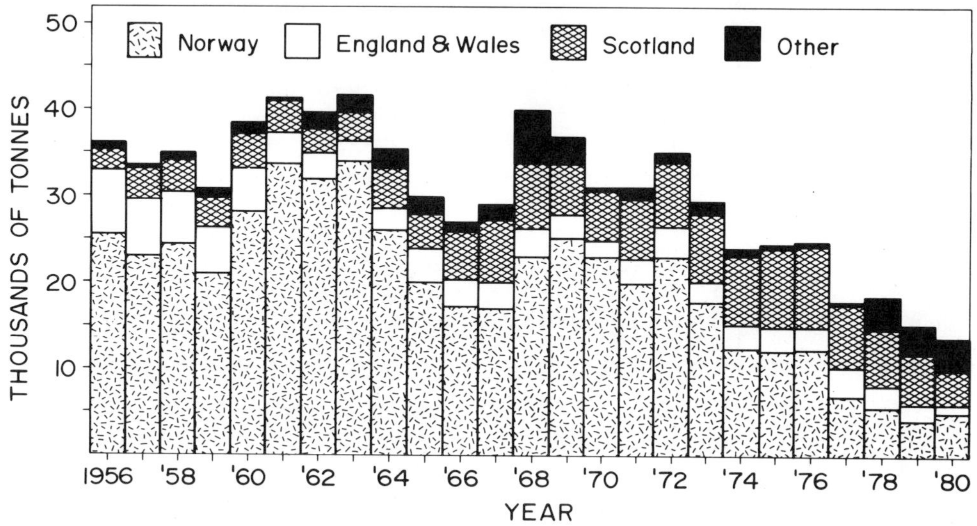

FIG 25. Long-term yield from the Scottish–Norwegian "picked" dogfish stock (Data from Holden 1977 and Bulletin Statistique des Pêches Maritimes. Cons. Int. Explor. de la Mer, Vol. 46–65).

TABLE 20. Washington State production of spiny dogfish by gear in Area 4A: Puget Sound and vicinity (figures in tonnes)[a].

Gear	1976	1977	1978	1979	1980	1981[b]	1982[b]
Trawl	532.2	523.7	677.8	1053.2	736.8	828.7	1143.0
Set-net	1846.8	1504.5	1185.3	1119.3	806.3	—	—
Set-line	198.1	332.5	779.5	1701.0	1455.8	525.0	390.0
Other	5.4	5.4	4.5	9.3	5.1	455.0	411.0
Total	2582.5	2366.1	2647.1	3882.9	3004.0	1808.7	1944.0

[a]Source: Appendix 11, based on Pedersen and DiDonato (1982: tables 8, 15, 22, 27, and 33).
[b]Preliminary data from unpublished reports submitted by Wash. State Dep. to 1981 and 1982 annual meeting of the Canada/United States Groundfish Committee. The category "Other" may include some set-net catches.

a) Events in the State of Washington

The year 1975 marked the small beginning of a food-fish fishery for dogfish in Washington with a landing of 188 t by trawlers (Pattie and Gormley 1977:6), 18 t by set-liners and little more than half a tonne taken under special permit by sunken gillnet (set-net) vessels fishing in Puget Sound (Pedersen 1980). Along with fish used for reduction purposes in that year the total landing apparently reached 508 t (Anon. 1981:31). However, the following year, 1976, was one of major development when production leapt to 2635 t (Table 12) of which 2583 t were taken in Puget Sound, primarily (72%) with sunken gill nets (Table 20).

Washington fishermen met with substantially better results using set-nets than did their Canadian counterparts 2 years earlier (p. 49), probably because the nets were designed specifically for catching dogfish and made of tarred heavy twine of 7 and 7½ inch (18–19 cm) mesh size (Pedersen 1980). This gear was selective for large sized dogfish and appropriate for meeting the minimum market size of 32 inches (81 cm).

As shown in Table 20 the landings of dogfish taken with set-nets declined from 1976 through 1981. Pedersen (1981) noted a number of reasons: (1) the abundance of fish larger than 81 cm, for which set-nets were particularly selective appeared to be declining, (2) set-line fishing was becoming more appealing because of the lower capital investment requirement, fewer time-area restrictions on fishing, less gear loss and besides, markets were accepting sizes of dogfish down to 26 inches (71 cm). In a limited comparison, Pedersen (1981) observed that set-lines were actually more selective for large dogfish than were set-nets which in turn were much more selective than trawls.

By 1981 the set-net share from Washington waters as a whole had fallen to 1.4%, while set-line landings increased to 27% with trawls and miscellaneous other gear accounting for 48 and 23%, respectively (Anon. 1981).

Landings by Washington vessels of all types rose to a peak of 4284 t in 1979 and thereafter began a steep decline to less than half that figure (2032 t) by 1982 (Table 12). Almost all of the production has come from Puget Sound or Area 4A (cf. Table 13) and within that area during the 1976–80 period, the North Sound and Straits-Admiralty subareas (contiguous to Canadian waters) accounted for nearly 70% (Table 21).

TABLE 21. Average annual landing of spiny dogfish by Washington vessels from subareas of Puget Sound and vicinity (Area 4A) for the period 1976–80 (figures in tonnes)[a].

Subarea[b]	Average landing	%
North Sound (80)	1225.2	43.3
Straits-Admiralty (86,89)	758.6	26.2
Hood Canal (88)	326.5	11.3
Central Sound (83,84)	434.0	15.0
South Sound (85)	120.1	4.2
Average total landing	2894.4	100.0

[a]Source: Appendix 11.
[b]See Fig. 3 for subareas.

While the landings from all Washington waters declined from 4284 to 2032 t between 1979 and 1982, the amount of product exported to Europe did not decline proportionately because substantial quantities of whole dogfish were being received from British Columbia for reasons given below. The total amount of dogfish processed in Washington in 1982 was estimated at 4146 t or only 30% less than in the peak year of 1979.

b) Events in British Columbia

It was not until 1977 that a significant fishery for dogfish, as food for human consumption, became firmly established in British Columbia. Initially, as in Washington, the market required fish no less than 32 inches (81 cm) in total length, and, because quality of product depended on the shortest possible time between capture and processing, the Strait of Georgia received the brunt of the fishing effort. Production in that year rose to 1730 t and 95% of this originated from the Strait. Catches by set-lines predominated (Table 22, Fig. 26) and were mainly from the generally untrawlable subareas along the mainland side of the Strait (Minor Statistical Areas, MSA 15 and 16 in Fig. 3).

Peak production from the Strait occurred in 1979, as in Washington, when 4334 t (91% of the British Columbia total) were landed. By that time 80% of the catch was being made by set-liners but MSA 17 on the Vancouver Island side of the Strait had become the main producer (Fig. 26).

In 1979 the number of companies engaged in processing or at least acting as brokers for sales to other Canadian producers rose to 9, as opposed to 3 in 1976, with one Vancouver company accounting for 77% of the receipts from set-liners and trawlers. In the following year (1980) when 11 companies were involved, British Columbia production fell slightly to 4544 t but the Strait of Georgia now accounted for only 46% (2104 t) (Table 22). This was the first clear indication of a decline in supply of fish in the Strait, although it was apparent at least a year earlier that set-liners were able to keep up their production only by moving to other areas in the Strait (Fig. 26). The decline occurred despite the fact that early in 1979 the minimum market length on dogfish had been reduced to 28 inches (71 cm). This inducement perhaps was offset by a differential price structure that emerged in mid-year wherein medium sized fish (less than 32 inches) brought only half the price of large fish (Fig. 27).

By 1980, trawler production had also declined and the major site of fishing had shifted to grounds off the west coast of Vancouver Island (Table 23), a region not particularly noted historically for the production of fish of large size.

In 1981 the Strait of Georgia share increased to 60% but by then production from that region stood at 1210 t or only 28% of that in the peak year of 1979 (Table 22).

Gross information on the catch and effort of the set-line fishery in the Strait failed to confirm the other indications of decline in supply. Estimated average CPUE (tonnes per 1000 hooks) showed no decline between 1979 and 1981 (Table 24). This may have been due to inaccuracy of information on numbers of hooks fished per trip; to the fact that the geographical and seasonal distribution of fishing was not uniform and therefore not comparable from year to year; and to the possibility that only the more experienced and capable fishermen remained in the fishery.

On the other hand, statistics of catch and effort for the Canadian trawl fishery in the southern part of the Strait of Georgia (Table 25) showed a decline in CPUE of roughly 70% between 1977 and 1982. Whether the magnitude of this drop in fishing success is a reliable reflection of the change in relative abundance of spiny dogfish and is representative of the rest of the Strait of

TABLE 22. Canadian production of spiny dogfish by gear in the Strait of Georgia and vicinity, 1976–82 (weights in tonnes)[a].

Gear	1976	1977	1978	1979	1980	1981[b]	1982[b]
Trawl	81.9	651.4	722.0	884.2	564.5	446.6	695.9
Set-line	88.5	923.5	2107.4	3450.0	1536.6	765.7	1314.7
Other	68.9	62.1	3.2	—	3.2	—	—
Total	239.3	1637.0	2830.6	4334.2	2104.3	1212.3	2010.6

[a]Source: Appendix 10.
[b]Includes estimate of amount of dogfish landed directly by Canadian vessels in the State of Washington — 449 t in 1981; 755 t in 1982.

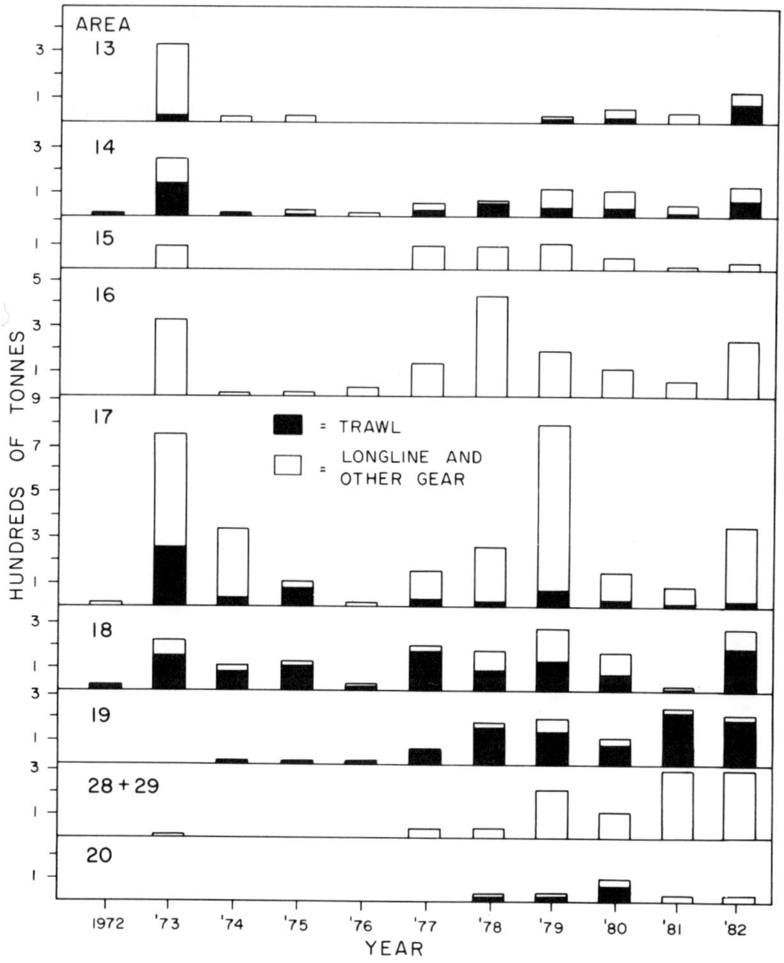

FIG 26. Canadian landings of spiny dogfish by gear, by minor statistical areas of the Strait of Georgia, 1972–82.

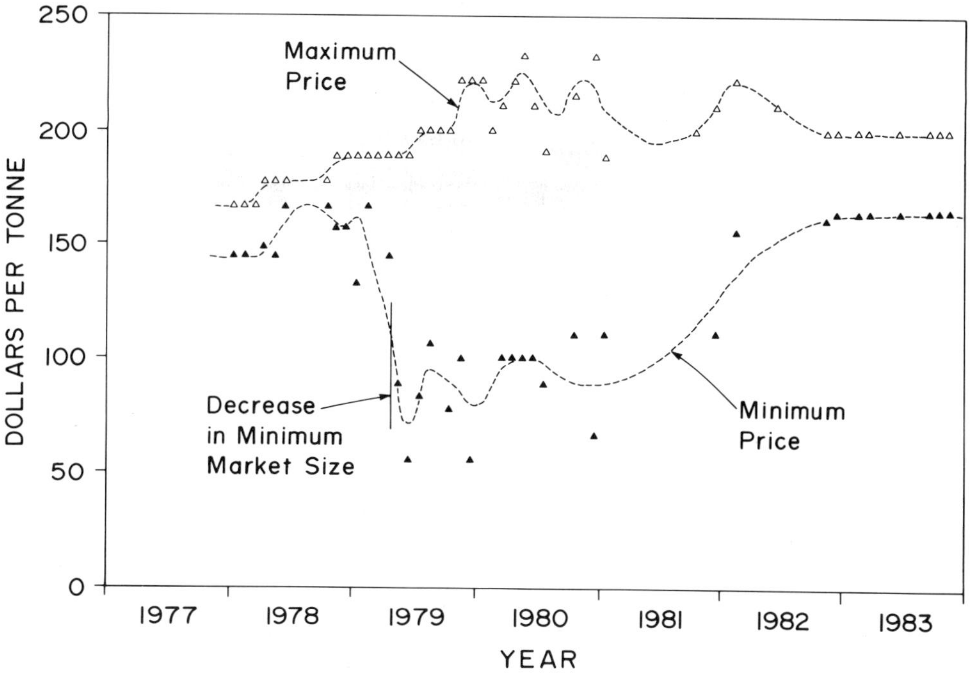

FIG 27. Minimum and maximum prices paid for spiny dogfish at Canadian south coast ports. (Source: British Columbia Fish Marketing Reports, Dep. Fisheries and Oceans, Vancouver, B.C.)

TABLE 23. Landings of spiny dogfish from grounds along the open coast of North America from northern California to Hecate Strait (figures in tonnes)[a].

Region	Statistical areas	California to Oregon 1C to 2C	Oregon to Washington 2D to 3B	Vancouver Island 3C to 3D	Queen Charlotte Sound 5A to 5B	Hecate Strait 5C to 5D	Total all areas
1976	U.S.	—	5	9	19	—	
	Can.	—	8	—	—	3	
	Total	—	13	9	19	3	44
1977	U.S.	2	136	41	—	—	
	Can.	—	—	78	6	11	
	Total	2	136	118	6	11	273
1978	U.S.	—	127	11	20	—	
	Can.	—	—	235	29	32	
	Total	—	127	246	49	32	454
1979	U.S.	4	388	2	—	—	
	Can.	—	—	299	15	109	
	Total	4	388	301	15	109	817
1980	U.S.	6	170	63	—	—	
	Can.	—	—	1872	161	405	
	Total	6	170	1935	161	405	2677
1981	U.S.	1	238	42	97	—	
	Can.	—	—	494[a]	25	52	
	Total	1	238	536	122	52	949
1982	U.S.	?	115	9	—	—	
	Can.	—	—	1559[b]	69	276	
	Total	?	115	1568	69	276	2028

[a]Source: Pacific Marine Fisheries Commission (1976–81); Leaman (1983) and unpublished records of Wash. Dep. of Fisheries (1982) and Oregon Department of Fish and Wildlife (1982).
[b]Includes adjustment for landings not officially reported.

TABLE 24. Statistics of the Canadian set-line fishery in the Strait of Georgia and vicinity (Area 4B)[a].

Year	Total catch[b] (t)	Monitored catch (t)	Monitored effort (Hooks × 1000)	Estimated total effort (Hooks × 1000)	Estimated average CPUE (t/1000) hooks
1979	3566.68	1696.62	2742.86	5766.11	0.619
1980	1599.95	606.77	923.01	2433.82	0.657
1981	544.16	379.92	516.14	739.27	0.736
1982	882.54	433.16	634.53	1292.82	0.683

[a]Source: Smith (1980, 1981); Leaman (1982); unpubl. data (1982).
[b]Primarily dogfish. Includes small amounts of incidentally caught species (rockfish, lingcod, Pacific cod, sablefish, etc.), usually less than 5%.

TABLE 25. Statistics of the Canadian trawl fishery for spiny dogfish in the southern part of the Strait of Georgia (Areas 4B; Minor statistical areas 17–19 incl.)[a].

Year	Qualifying catch[b] (t)	Fishing effort[c] (h)	Estimated average CPUE (t/h)
1977	487.30	772.5	0.631
1978	371.98	706.2	0.527
1979	255.32	655.9	0.389
1980	137.14	621.4	0.221
1981	99.99	487.6	0.176
1982	23.59	147.0	0.160

[a]Source: computer records maintained by DFO Fisheries Research Branch, Pacific Biological Station, Groundfish Investigation.
[b]Records of only those vessels for which fishing effort was reported and whose landings of spiny dogfish represented at least 25% of total weight of fish landed from each trip.
[c]Fishing effort of only those vessels whose landings met the 25% catch qualification. This restriction attempts to identify fishing effort that was directed specifically to the capture of spiny dogfish.

Georgia, including contiguous U.S. waters, remains uncertain. There is some evidence of a progressive change in the geographical pattern of herring spawning and this may have affected the availability of dogfish to trawls.

Further indications of a decline in the abundance of spiny dogfish in the Strait of Georgia is suggested by the shrinking differential in pricing of medium and large dogfish after 1980 (Fig. 27). Prices for medium sized dogfish increased relative to those of large fish, probably reflecting efforts of buyers to encourage continuing supply.

It was obvious, however, that in addition to the problem of declining supply and hence declining income for fishermen, the Canadian industry was beset by other financial difficulties. From the start of the fishery in 1977 a substantial fraction of spiny dogfish landings that could not be processed profitably in British Columbia was each year shipped in the round (fresh) by truck transport or delivered directly by packers and fishing vessels to plants in the State of Washington. Throughout the 1977–82 period just over half of the spiny dogfish caught by British Columbia fishermen was exported in this unprocessed form (Table 26). Along with it went employment for shoreworkers and income that might otherwise have remained in Canada. Had all of the fish caught been processed in British Columbia for markets in Europe and the Orient, the wholesale value of Canadian landings could have been increased by more than three-quarters of a million dollars per year.

One would expect that the British Columbia industry has a competitive advantage over that in Washington State in supplying foreign markets if only because of the more than 20% currency differential on the dollar. However this advantage appears to be more than offset by the relatively high costs of shore handling and processing. In addition to high wage demands, employers are required to contribute to a number of social benefits (workers compensation, unemployment insurance etc.) — costs not faced by their counterparts in the State of Washington. On the other hand, it is not unreasonable to suggest that employers who derive their profit mainly from the lucrative salmon and herring fisheries of British Columbia are unlikely to have much enthusiasm for developing processing industries (and seeking markets) for a marginally profitable enterprise such as that on spiny dogfish.

c) Summary of Inshore Events

Although there has been some fishing activity in all areas of the British Columbia coast in recent years (Table 23) most of the Canadian fishery has been concentrated in the Strait of Georgia. The American fishery likewise has been confined largely to contiguous internal waters of Washington. Since we have already established (Chap. 3, pt. 4) that there is at least some intermingling of spiny dogfish between these inshore waters (Areas 4A and 4B) and that the inshore stock(s) appear to be somewhat separate from the stock(s) along the open coast of North America, it is appropriate to summarize events of the continuing foodfish era in Washington and British Columbia by considering the combined production figures for the two inshore fisheries.

We have set 1976 as the starting date of the foodfish fishery, but it was in fact preceded by a trial (government assisted) fishery in 1973 and 1974 and some marginally effective private ventures in 1975 (Fig. 28). Landings rose to 2800 t in 1976 and still further to a peak of 8200 t in 1979 but then declined to only half that level by 1982. (Latest information suggests that there was no improvement in this situation in 1983.) We are in a poor position

TABLE 26. Exports of spiny dogfish and processed products from British Columbia (quantities in thousands of kilograms and wholesale value in thousands of dollars)[a].

Product		1977	1978	1979	1980	1981	1982	Destination
RAW								
Round (fresh)[b]	Q	902	1882	1707	2165	1092	2114	USA
	V	119	332	301	382	214	524	
PROCESSED								
Dressed (frozen)	Q	—	77	205	187	97	11	Japan & USA
	V	—	81	215	121	38	4	
Belly-flaps	Q	69	81	182	131	21	108	Europe
	V	163	192	431	322	60	247	
Backs	Q	218	259	567	617	207	521	Europe
	V	250	297	651	690	208	546	
Fins and tails	Q	—	—	—	40	7	32	Orient
	V	—	—	—	41	8	36	
Raw product as % of total B.C. landing		52	60	36	48	61	54	

[a]Source: Statistics developed by the writer on the basis of a special investigation of published and unpublished records of the Department of Fisheries and Oceans, Economics and Statistical Services Branch; statistics produced by the Province of British Columbia, Marine Resources Branch; Fisheries Market News published by the National Marine Fisheries Service (Seattle).

[b]Includes small quantities of dressed fresh fish converted to round weight.

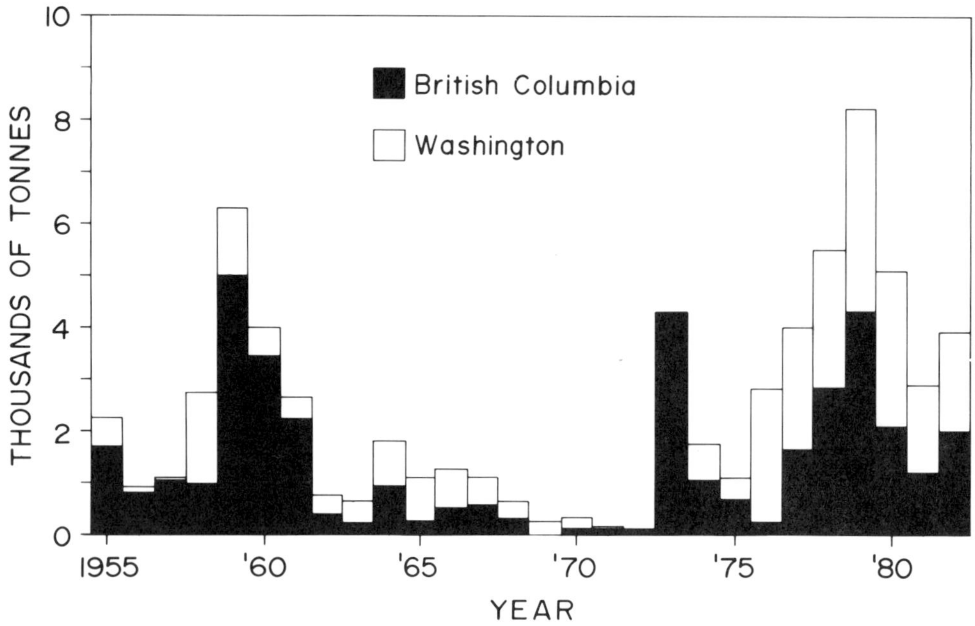

FIG 28. Production of spiny dogfish from internal waters of Washington (Area 4A) and British Columbia (Area 4B), 1955–82.

to supply an unequivocal explanation for the decline — i.e., whether it was due primarily to a drop in abundance of spiny dogfish or to general economic difficulties. Statistics of catch and fishing effort, traditional tools for detecting changes in fish abundance, have not been universally revealing. Records of the Canadian trawl fishery indicate a significant decline, but those of the set-line fishery do not. Yet we have reasonable grounds for saying that even before 1979 the abundance of large fish (81 + cm) had begun to decline. Otherwise, why did the processors relax their minimum size limit in mid-1979? In Washington, abandonment of the use of set-nets (highly selective for large fish) was in part due to reduction in abundance. The shift of Canadian trawlers to waters off the west coast of Vancouver Island was likewise indicative of declining productivity inshore. Why the Washington trawler fleet did not follow suit is not clear. One possible explanation is that grounds off the open coast of Washington are less productive of dogfish than those on La Pérouse Bank, lying within Canada's zone of extended fishery jurisdiction.

However, it would be naive to assume that overfishing alone was responsible for the decline in landings from inshore waters. Marginally profitable fisheries are acutely sensitive to market demand and can be severely affected by developments in other areas. For example, there appears to be increasing competition for European markets from processors on the east coast of the United States and Canada. There, costs of labor and shipping to Europe are substantially lower than in the State of Washington, to say nothing of British Columbia where shoreworkers' wages are the highest in the world. Statistical bulletins of the Northwest Atlantic Fisheries Organization (NAFO) show a rise in American landings of spiny dogfish from less than 1000 t in 1977 to nearly 7000 t in 1981. Landings from waters off the Atlantic provinces of Canada where the species is a seasonal visitor have been of a lesser amount and more variable. After rising to a peak of 1000 t in 1979 (under subsidy), production sank to less than 600 t in 1981.

Most of the yield of spiny dogfish from the Northwest Atlantic is destined for the same European markets supplied from the west coast of Canada and the USA. However, some products are now finding acceptance on the fresh fish market of New York City (B. Skud, NMFS, Rhode Island, personal communication).

d) Management Considerations

To date no action has been taken by Canadian and American fishery managers to limit the catch of spiny dogfish in the Strait of Georgia, Puget Sound and adjoining inshore waters. However biologists of both countries have developed tentative management plans for groundfish which include provisional recommendations on catch limits. For areas under the jurisdiction of the State of Washington, Pedersen and DiDonato (1982) have defined maximum sustainable yield (MSY) as ". . .the average over a period of years of the largest catch which can be taken continuously from a stock"[2]. They have also defined an acceptable biological catch (ABC) as an annually determined figure which, for biological reasons (e.g., year-to-year variations in recruitment), may be larger or smaller than the MSY. For spiny dogfish,

[2]The term MSY has received much criticism in scientific literature chiefly on grounds that it denotes an objective that, in its strict sense, is difficult if not impossible to attain. As used here it appears to have a looser definition — an approximation akin to Gulland's (1983) view that it is not so much an objective to be rigidly followed in reaching a decision but rather a convenient concept to be used in discussing general management problems.

Pedersen and DiDonato (1982) have proposed an MSY and ABC of about 2900 t, this being nothing more than an average of the Washington catch during the period 1976–80. Clearly this is a tenuous estimate but in the absence of a data-base on changes in abundance or an estimation of biomass there appears to be no better guideline at present time. A "fishing up" of reserves that had accumulated during the 20 years prior to the foodfish fishery may have exaggerated the apparent abundance of dogfish, in which case MSY is probably overestimated.

For those parts of the Strait of Georgia under Canadian jurisdiction, the basis for calculating MSY, though seemingly more sophisticated, is no less precarious. Wood et al. (1979) estimated that a spiny dogfish stock of 200 000 t occupied waters off British Columbia (inshore waters included) prior to the great liver fishery of the 1940s, and went on to calculate an MSY of 8000–10 000 t using Gulland's (1970) approximation. The MSY for the Strait of Georgia was simply assumed to be about one-third of the total for the British Columbia coast, namely, 3000 t (Stocker 1981). The recommended total allowable catch (TAC) was an equal amount or, about the same as the ABC selected by Washington biologists for their inshore waters. Canadian managers have been prepared for several years to act on this guideline, but since 1979, the catch has never approached the 3000 t level.

In any event coordination of Canadian and American management plans appears to be essential in view of the evidence from tagging results of intermingling between the two areas of jurisdiction. Further comment on the problems and prospects of managing fisheries for spiny dogfish has been reserved for the next chapter.

Chapter V. The Future

In this concluding chapter we shall examine some of the economic prospects for continuation of the foodfishery for spiny dogfish in the Northeast Pacific. Also deserving consideration are the prospects for development of a management regime that will not only assure continuation of such a fishery but also satisfy concerns about the species' interference with other fisheries and alleged depredation of stocks of other, more valuable species. The extent to which current and yet-to-be-acquired scientific knowledge can shed light on these issues will be examined also.

1. Industrial Prospects

It is perhaps no easier now than it was 50 to 100 years ago to foretell events in the Northeast Pacific fishery for spiny dogfish, although the recently evolved market as a foodfish seems conducive to greater stability. The fact that now, as during most of its history, the industry operates on a narrow margin of profit, certainly assures that there will be on- and off-periods of activity related to the general economic climate, fluctuations in exchange rates, cost of fuel, labor unrest — all quite aside from the state of the spiny dogfish stocks themselves.

Continued participation in the European market may depend on whether Northwest Atlantic production rises to a level where it alone satisfies the import requirements. That being the case, there remains the potentially large market in the Pacific rim countries of Asia, access to which is currently inhibited by processing and shipping costs (except in the case of the market for fins and tails). In any event, dependence on foreign markets, particularly those that are vyed for by more than one national supplier, will always be plagued with uncertainty.

Significant development of a stable domestic (Canada–USA) market seems a long way off, notwithstanding developments in New York City. North Americans and particularly Canadians (not noted as fish eaters even under the best of circumstances) would have to be educated to accept the lowly dogfish as a food. That process could be hastened only if the price of the currently least expensive seafood products were to rise beyond the reach of the average householder.

At time of writing, the large and semi-migratory stock(s) of spiny dogfish in Hecate Strait and off the open coast of North America from the Queen Charlotte Islands to northern California have remained relatively untouched by the recent flurry of foodfish production. The only area of significant activity is that in the Canadian zone along the west coast of Vancouver Island (Table 23), and, as already noted, this appears to be symptomatic of declining fishing success in inshore waters. The expansion to seaward is largely a Canadian phenomenon. Vessels fishing there make their landings either in Vancouver, B.C. or Bellingham, Washington.

Because of the necessity to market spiny dogfish in a fresh condition, the distance that a fleet can operate from port, that is, the time spent away from port, is now much more limiting. The west coast of Vancouver Island grounds are attractive because their greater distance from port is offset by high catch per unit of effort and short "turn around" time.

Hecate Strait, the most promising area for development, on the other hand is largely beyond the working range of fresh-fish vessels based in Vancouver or elsewhere on the south coast. The Strait would be more accessible if dogfish processing operations at Prince Rupert or other northern British Columbia ports could be made more economical. However, costs of fishing and processing in the north are currently even higher than those in the Vancouver area and hence, for the time being at least, much too high to attract a purely domestic land-based fishery.

Processing at sea has been tried by a few Canadian vessels including a factory-trawler fishing in Hecate Strait but these ventures were short-lived, suggesting that this type of operation is precluded because of the high costs of capital equipment and labor.

Consideration has been given to allowing foreign factory-trawlers to exploit spiny dogfish in Canada's zone of extended jurisdiction. In 1977 an experimental licence was issued to a Polish vessel for this purpose but was soon revoked because of the difficulty of avoiding incidental catches of species that were already fully utilized by Canadian fishermen.

The most likely prospect is a joint venture such as the one now taking place on Pacific hake *(Merluccius productus)* off the southern British Columbia coast, wherein Canadian trawlers do the fishing and supply their catches to foreign processing vessels. It too poses a problem of incidental catches, but less so because of the likelihood that fishing could be conducted more selectively. For example, a sunken gill-net operation might be the most appropriate one for Hecate Strait. Such gear was used on spiny dogfish during the 1940s with great effectiveness and apparently with only minor incidental catches and a minimum of physical interferences with the set-line, trawl, and troll fisheries for halibut, other groundfish, and salmon, respectively.

These prospects exclude consideration of the future of the fishery as it now exists in the Strait of Georgia and Puget Sound — a future that is dependent not only on the abundance of spiny dogfish and marketing opportunities but also on the short- and long-term policies that the respective fishery management authorities choose to pursue. Before considering the various options it is instructive to examine what scientific evidence is available, recognizing that social and economic factors may play an equal if not greater role in the decision-making process.

2. The Foundation for a Management Policy

It would not be unreasonable to say that, if senior managers in the Canadian Department of Fisheries and Oceans and the Washington State Department of Fisheries were to articulate a policy for management of the spiny dogfish, its substance would be to promote commercial exploitation as a means of reducing and thereby controlling the influence of spiny dogfish on fisheries for other, more valuable species. Such a viewpoint stresses the welfare of species other than the spiny dogfish itself. It would owe its acceptability to the fact that, while relatively few fishermen and shoreworkers depend on spiny dogfish for a living, thousands of people employed in the commercial salmon, herring and halibut fisheries, to say nothing of a large and influential sport fishing fraternity, deplore the very existence of the spiny dogfish. Their condemnation is based on the reputation this animal has acquired as a nuisance or pest in respect to its interference with the prosecution of other fisheries and as a despoiler of other species. The validity of these accusations deserves closer scrutiny.

a) The Nuisance Problem

There is no difficulty in documenting at least qualitatively the extent of the spiny dogfish's interference with other fisheries and the damage it inflicts on fishing gear. The problem is worldwide and stems from the fact that the species is given to travelling in dense schools. When it makes its appearance on a particular ground it is usually in such large numbers as to make fishing for other species impossible. This is the most common form of interference and the present day set-line fishery for halibut in Hecate Strait provides a good example. There, depending on the locality, stage of tide and time of day, it is difficult if not impossible to catch halibut before the baited hooks have been pre-empted by dogfish. This constitutes a waste of manpower and bait, and on occasion contributes to delay in taking the allowable annual catch. Trawlers, although they usually try to avoid spiny dogfish, at times catch such large numbers that hours are spent clearing the nets, sometimes at risk to life and limb, and costly damage to rigging.

Gear damage takes the form of cut or abraided lines and lost hooks; torn or frayed meshes in gill nets, seines, and herring impoundments; broken hauling equipment and other rigging, as well as damage to other species present in the catches — all the result of the spiny dogfish's sharp teeth, spines, and sandpaper-like skin.

Placing a monetary value on such losses is not an impossible task. Indeed, we have already reviewed (p. 44) some calculations of damage in years past to fisheries in the State of Washington. While such estimates may be subject to exaggeration when not made by disinterested observers, they are nevertheless obtainable. Indeed they are essential for any cost/benefit analysis of a control program proposed for any area where no significant effort is being made to utilize spiny dogfish commercially.

b) The Predation Problem

Much more difficult to find is irrefutable, quantitative evidence of the mortality that spiny dogfish inflict on other species, particularly on salmon and herring. The controversy regarding the impact on salmon has raged for decades and grossly misleading remarks such as the following are commonplace:

> "Although he [the spiny dogfish] enjoys such a variety of food, evidently the particular delicacy for this cannibal creature is salmon, which he will devour at any stage of its growth from young salmon newly arrived in the ocean to mature individuals making the return migration." (Lyons 1969:582).

While there is no question that spiny dogfish are quick to take advantage of salmon entangled in nets, caught on lures, injured and briefly disoriented after escaping or being released from nets or lures, the fact is there is little evidence that the strong but slow-swimming dogfish can and does consume salmon in their natural, unhampered state. The spiny dogfish has an active metabolic rate scarcely more than one-sixth that of an adult salmon (Brett and Blackburn 1978). Of more significance, however, is the fact that its body is simply not built for speed. It has an "anguilliform" (eel-like) mode of locomotion, as opposed to that of salmonids which tend toward the "carangiform" mode typified by the tunas and mackerels (Webb 1975). In short it would be an insult to a healthy salmon to say that it would have any difficulty in out-manoeuvering a pursuing dogfish.

On only rare occasions have fingerling salmon been found in the stomachs of spiny dogfish. Their feeding on newly emerged chum fry at the mouth of a small stream was observed on one occasion (J. C. Mason, Pac. Biol. Stn., Nanaimo, personal communication) but whether this is of common occurrence is not known. A recent study of marine predators off the estuary of the Big Qualicum River, at the time of seaward migration of pink and chum fry, revealed the culprits to be not spiny dogfish but adult and large juvenile chinook and coho salmon (Robinson et al. 1982).

Although the spiny dogfish is a highly opportunistic feeder and has a widely varied diet, there is no question that herring are an important item of food in certain areas and at certain times of year. However, there is serious doubt that predation by dogfish alone is of such magnitude in British Columbia waters as to limit significantly the supply of herring to commercial fishermen. Jones and Geen (1977), the first and only researchers so far to tackle this difficult problem concluded that spiny dogfish consume 230 000 t of herring annually, that is, an amount nearly equal to the all-time-record annual harvest.

However, for a number of reasons this estimate must be called in question. In the first place the basis for the authors' calculations was severely distorted by the unrepresentativeness of the stomach data available to them. Of nearly 18 000 spiny dogfish examined, 64% were from the Strait of Georgia (Area 4B) with 24% from the southwest coast of Vancouver Island (Area 3C). No stomach samples were available from the adjacent Area 3D and only 2143 (12%) were obtained from all Queen Charlotte Sound (Areas 5A and 5B) and the major dogfish area of Hecate Strait (Areas 5C and 5D). These figures are in no way proportional to the distribution of herring or of dogfish.

Hourston and Haegele (1980) estimated the adult population of British Columbia herring to be about 350 000 t in the middle to late 1970s. More recent studies have led to a downward revision of this figure to 300 000 t

(D. Ware, Pac. Biol. Stn., Nanaimo, personal communication). Applying the best available estimates of annual mortality rates (36% due to natural causes; 18% due to fishing) to fish from the beginning of their third birthday to end of life, the estimate of natural loss comes to only 108 000 t. Yet this figure represents deaths from *all* causes: predation by at least 15 common fish species (in addition to spiny dogfish), numerous species of sea birds, seals, sea lions, killer whales, as well as losses due to disease and parasitism. Even when we make allowances for some dogfish predation on the juvenile and immature herring that do not participate in the seasonal inshore–offshore migration of spawners, it is difficult to conceive of dogfish predation alone amounting to anything over 80 000 t annually.

Finally, there is nothing in the historical record of the British Columbia herring stocks to suggest that the species flourished during the 1940s and early 1950s when spiny dogfish were in obviously low abundance. Of course, it can be argued that, because spiny dogfish eat other species that also eat herring, these other species increase in abundance when dogfish decline, and eat just as much herring as if dogfish were still abundant. This has been a matter of conjecture for at least 40 years and it seems destined to remain so for another 40 years or for as long as the spiny dogfish receives low priority in fisheries research.

In this and the preceding section we have reviewed as far as is currently possible the grounds for claiming that the spiny dogfish does so much damage to gear, interferes with fishing operations so badly and decimates stocks of more valuable species to such an extent that a "pest control" program would be of unquestionable benefit to the British Columbia, Washington and other state fisheries. The damage and interference aspect of the problem in British Columbia waters has never been sufficiently well documented in monetary terms to permit a reliable cost/benefit analysis. Although it could be done given sufficient resources, it seems unlikely that damage and lost wages alone would come anywhere near the cost of mounting an effective program. Adding the predation aspect does little to help the case because, as we have seen, the problem is shrouded in uncertainty and controversy. Much too little effort has been spent in trying to understand and measure the complex feeding behavior of spiny dogfish.

Thus the value of control measures (contract killing, bounty payments, etc.) is difficult to estimate. Certainly the Canadian experience of the late 1950s and early 1960s with government-subsidized control programs was sufficient to demonstrate the heavy cost to taxpayers and the dubious results of localized eradication efforts. Our increased knowledge of the movements and migrations of dogfish reveals the international scope of the problem and the difficulty that would arise in determining equitable national shares of cost and benefit. The fact that in the 1940s a removal of over 200 000 t was required to make an appreciable yet only temporary dent in the size of spiny dogfish stocks in the Northeast Pacific emphasizes further the enormity of the task.

Future funding of a control program, if it were ever again to be considered either by government or by the potential beneficiaries within the industry, is likely to be spasmodic. Certainly governments are reluctant to commit funds from one fiscal year to the next and, of course, quite unable to commit beyond their projected term of office. Thus if funds were to be allocated they would apply to brief intervals which would in turn lead to "pulsed" fishing operations.

The wisdom of such actions has recently been called into question. Indeed, there is now fairly convincing theoretical evidence that "pulsed" fishing, looked at over a span of several decades, is likely to impart such instability to spiny dogfish stocks as to aggravate the problem of management (Wood et al. 1979). This brings us to the only plausible alternative to damage- and predator-control programs.

c) The merits of a sustained fishery

Until the last decade most of the mathematical models used to assess the effects of fishing on stock production assumed that the rate of natural increase responds immediately to changes in population density and that the rate of natural increase at a given density is independent of age composition of the population. Such assumptions, as first noted by Holden (1977), do not apply in the case of the spiny dogfish. The reason is that the species is characterized by exceptional longevity, slow growth and a low reproductive capacity which in turn results from late maturity, low fecundity and a long gestation period. While these rather unique processes have been known for many years, it was not until methods were developed to provide reasonably accurate estimates of age and growth rate that it became possible to identify age at recruitment, age at maturity and the change in fecundity with age. With this information at hand, along with an estimate of the natural mortality rate, it then became possible to develop a mathematical, age-structured model that could be used to test various hypotheses about the natural mechanism that limits a dogfish population's growth and to determine how it operates in response to the effect of fishing. By comparing various simulated population changes with those actually observed it then becomes possible to determine which hypothesis is most compatible with the facts.

Wood et al. (1979) developed such a model and concluded that there is a density-dependent compensatory change in the rate of natural mortality, namely, that the natural death rate at all ages decreases as abundance is reduced by exploitation and increases as the population rebuilds to its primitive (natural) level of equilibrium. Just how the mortality rate is reduced, whether by diminished cannibalism, predation, competition, disease or parasitism, remains uncertain, since there are few clues from field observation to favor any of these possibilities. This uncertainty serves to remind us of the provisional nature of the model and the likelihood that, because of an inadequate data base, some of the underlying assumptions are oversimplified.

Nevertheless, support for the hypothesis of a density-dependent response is found in the ability of the model to predict observed patterns in simulation of the historical fishery. For example, as shown in Fig. 29, it predicted (in hindsight) the surprising resurgence of the "marketable stock" which occurred during the late 1950s and which we have already documented in a qualitative way on p. 44 et seq. Further support for the model is to be

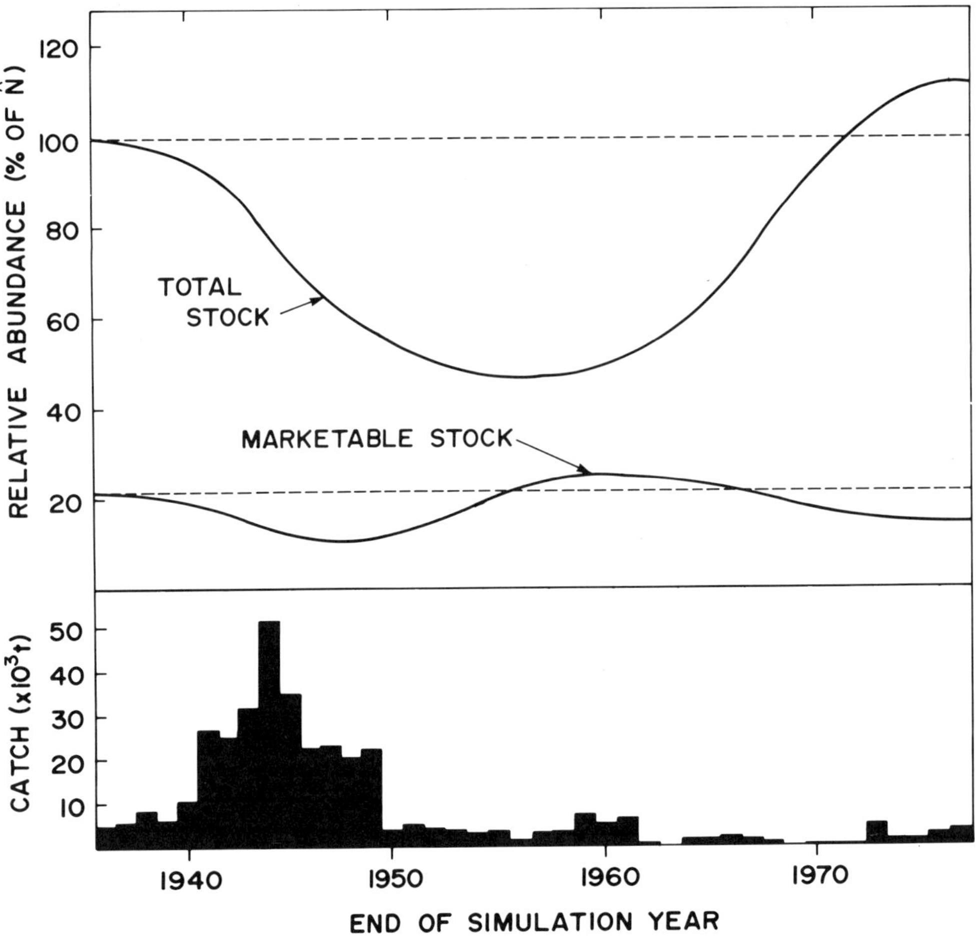

FIG 29. Trends in abundance of a "British Columbia" stock of spiny dogfish predicted by simulation of the historical fishery based on a compensatory mortality model — adapted from Wood et al. (1979; fig. 2). The degree of compensatory mortality employed is intermediate in a range of plausible values considered by those authors. Although this particular model simulates a stock with both inshore and offshore components, its performance differs only in detail from models designed to simulate the history of fishing on either an inshore or an offshore stock separately.

found in the favorable correspondence between predicted and observed changes in age composition and litter size (for further details see Wood et al. 1979).

At time of writing, attempts to use this model to simulate the history of the spiny dogfish fishery in inshore waters — a history quite different from that along the open coast or for the whole Northeast Pacific — have been hindered by incomplete or missing information. One question, only partly answered, concerns the fundamental issue of stock discrimination: the true extent of intermingling between inshore Areas 4A and 4B and the extent to which spiny dogfish populations in these areas gain recruits from, or lose them to, stock(s) along the open coast. To be sure, there is qualitative evidence of an interchange but a clearer picture is necessary for reliable analysis.

A more important question arises from the lack of a reliable, absolute estimate of marketable-stock size at some stage during the history of the inshore fishery, preferably since the inception of the foodfish fishery. Alternatively, or better in addition, a measure of the rate of change in stock size (a relative index) during the course of that fishery is needed along with data on age composition of the landings. To date, Canadian catch statistics of neither the set-line nor trawl fisheries in inshore waters have provided trustworthy indices, and representative sampling of catches has posed some vexatious though not insurmountable problems.

These and other items of information, simple to perceive but difficult to obtain as long as research priority on spiny dogfish remains low, would provide a point of entry to the Wood et al. (1979) model and permit us to see which set of population parameters is most compatible with events in the real world. Having done so, it should then be possible to determine the most appropriate sustainable level of catch. More importantly, the consequences of various management strategies would be testable.

In their preliminary assessment of strategies for exploitation of spiny dogfish stocks that are at or near their natural level of equilibrium, Wood et al. (1979) recommended cautious development of a sustained-effort fishery. Such a condition may be difficult to fulfill since markets often develop rapidly and sustained effort may be less easy to control than sustained catch (e.g. a quota). The latter is theoretically less acceptable because it imparts oscillations to the stock level. However, they would be of such long period and low amplitude that they would likely escape detection by methods currently available for measuring changes in abundance (see Wood et al. 1979; Fig. 6). Thus from a practical standpoint a sustained yield policy would not be measurably less desirable than one of sustained effort.

The recommendation of a cautiously developed fishery stems from the fact that a previously unexploited population of spiny dogfish is exceptionally sensitive to small increases in fishing mortality. If development is too rapid a chain-reaction in the form of long-term cycles in abundance is set in motion. A similar but even more violent, and hence less manageable, response is likely to result from intensive periodic or "pulsed" fishing such as we have already mentioned.

This predisposition to population instability stems from the long time-lag between reproduction and recruitment. Referring again to Figure 29, if we were to assume that instead of a short but intense fishery for dogfish liver in the 1940s there had been a costly eradication or "pest control" fishery, it would have had to be repeated around about 1960. We are presuming here that fish in the marketable stock are the ones that cause the most disruption of other fisheries. This is certainly true for the bottom set-line and trawl fisheries and perhaps less so for the pelagic gillnet and seine fisheries.

Another point of interest is the long duration of dogfish population fluctuations. If the time-scale of Fig. 29 were extended, we would find that oscillations created by the brief but intensive fishery in the 1940s are likely, at least in theory, to extend well into the 21st Century.

While recommendation of a slowly developed sustained-yield fishery may be appropriate from the standpoint of the spiny dogfish resource, a number of important questions remain to be answered. Wood et al. (1979) estimated that the maximum sustainable yield (MSY) is attained when the marketable biomass of a spiny dogfish population is reduced to 57% of the unharvested level. The MSY itself is little more than 9% of the marketable biomass. For lack of detailed knowledge of the economics of the current foodfish fishery, we must leave unanswered the question of whether dogfish fishermen could afford to continue operating at such a level of stock. At current prices the success of their operation depends on a relatively high catch-per-unit-effort, quite possibly higher than that prevailing at an MSY stock level. In other words to achieve the maximum sustainable *economic* yield may require the maintenance of stocks at a level much higher than that which provides an MSY in terms of weight alone.

Even if an MSY stock level permitted a viable, continuing fishery for spiny dogfish, we are left with the need to determine whether this level satisfies the concerns of managers and the rest of the fishing industry. It is conceivable that a sustained-yield fishery is not in the best interest of the British Columbia and Washington fishery economy as a whole. Stock levels might still be too high to solve the interference and nuisance problems. Again, no complete answer is forthcoming. Yet it does seem clear that reduction of the marketable stock (mainly adults and large immature fish) to the MSY level will reduce measurably the interference factor for set-liners, trawlers and other vessels dependent on production of valuable species living on or near the ocean bottom, for this is where the marketable stock of spiny dogfish is most usually found.

The remainder of the unexploited stock (juveniles of unmarketable size) constitutes nearly 80% in terms of numbers but only about 40% of the total biomass and tends to live a more or less pelagic life. Its reduction, in response to a gradually developed sustained-yield fishery, will come about slowly over the long term. Thus for pelagic fisheries, the benefit of reduced interference may be difficult to measure.

As to the predation factor, it seems prudent that, until direct scientific investigation can provide reasonable evidence that resources of more valuable species can be restored to higher levels by suppression of spiny dogfish, action should be held in abeyance.

To sum up, scientific assessment of the options for managing and/or controlling the stocks of spiny dogfish in the Northeast Pacific points to the merits of a sustained fishery as opposed to a spasmodic one. Quite aside from the desirable stability that the former course of action would tend to create, a sustained fishery is one that is most compatible with the need for a sound economic operation: assured continuity of supply of the raw product. Whether a dogfish fishery could survive economically when stocks are exploited at or near their maximum sustained (physical) yield remains an unanswered question.

In any event a clearly viable commercial venture has much to commend itself, as opposed to *ad hoc* control programs funded by government or industry. The latter should be avoided until there is clear scientific evidence of their need.

While the basis for the various forecasts and calculations now emerging through sophisticated computer programming is in need of much more investigation, there is no question that biologists are in a better position than ever before to provide guidance in management of the spiny dogfish resource. Indeed, some future chonicler of events in the Northeast Pacific may be moved to call the mid-1980s the beginning of a "fifth era" in the long history of fishing for that species: one of scientific management in which future events stand some chance of becoming *planned* events.

Acknowledgments

I wish to express my appreciation for the helpful suggestions and access to unpublished data provided by G. A. McFarlane, biologist-in-charge of present day investigations of spiny dogfish at the Pacific Biological Station at Nanaimo, and by Mark G. Pedersen, Chief of the Groundfish Management Division of the Washington Department of Fisheries. Their advice in respect to the modern-day fishery has been most valuable, as was the assistance of the following staff members of the Pacific Biological Station: F. Crabbe, J. Leaman, G. Miller, M. Saunders, M. Smith and N. Venables; M. Kostner of the Economics and Statistical Services Branch, Vancouver.

I am also grateful to Prof. Knut R. Fladmark of the Department of Archaeology at Simon Fraser University and John Dewhirst of Parks, Canada, for their patient assistance in acquainting me with archaeological literature and its interpretation.

J. K. Pope, retired fisherman and three retired fishing company executives, F. W. Millerd, H. Mosdell, and D. R. Russell gave freely of their advice and recollections of the fishery in the 1930s and 1940s.

Finally I should like to thank a number of individuals who undertook to read the manuscript and who offered many helpful suggestions for its improvement: R. J. Childerhose, Z. Kabata, B. M. Leaman, G. A. McFarlane, J. A. Thomson, M. Waldichuk, S. J. Westrheim, and C. C. Wood.

References

AFS Committee on Names of Fishes. 1970. A list of common and scientific names of fishes from the United States and Canada, 3rd. ed. Am. Fish. Soc. Spec. Publ. 6: 150 p.

Alverson, D. L. 1958. Review of the dogfish *(Squalus acanthias)*, its utilization and its fishery. U.S. Dep. Inter. Fish Wildl. Serv. Bur. Commer. Fish. Mimeo Rep. 60 p.

Alverson, D. L., and M. E. Stansby. 1963. The spiny dogfish *(Squalus acanthias)* in the northeastern Pacific. U.S. Dep. Inter. Fish Wildl. Serv. Spec. Sci. Rep. Fish. 447: 25 p.

Alverson, D. L., A. T. Pruter, and L. R. Ronholt. 1964. A study of demersal fishes and fisheries of the northeastern Pacific Ocean. McMillan Lecture Ser., Inst. Fish. Univ. B.C. 190 p.

Anderson, A. C. 1872. Extract from government prize essay. Natural productions — Fish. Can. Dep. Mar. and Fish. Rep. for year ending 30th June 1871, App. R: 181–186.

1877. Report of the inspector of fisheries for British Columbia for the year of 1876. Can. Dep. Mar. and Fish. Ninth Ann. Rep., Rep. Commissioner of Fish., Suppl. 4, App. 21: 339–347.

1878. Report of the inspector of fisheries for British Columbia for the year of 1877. Can. Dep. Mar. and Fish., Tenth Ann. Rep., Rep. Commissioner of Fish., Suppl. 5, App. 17: 287–308.

1880. Report of the inspector of fisheries for British Columbia for the year 1879. Can. Dep. Mar. and Fish., Twelfth Ann. Rep., Rep. Commissioner of Fish., Suppl. 2, App. 17: 280–301.

1881. Report of the inspector of fisheries for British Columbia for 1881. Can. Dep. Mar. and Fish., Thirteenth Ann. Rep., Fish. Statements for the year 1880. Suppl. 2, App. 12: 257–267.

1884. Report of the inspector of fisheries for British Columbia for 1883. Can. Dep. Mar. and Fish. Sixteenth Ann. Rep.; Fish. Can., Suppl. 2, App. 7: 190–207.

Anon. 1892. Records of proceedings and minutes of evidence, etc. B.C. Salmon Fish. Comm., 431 p.

1893. Annual report of the Minister of Mines for the year ending 31st December 1892 in the Province of British Columbia: 521–636.

1944a. The rise of Pacific trawl fishing. Pac. Fisherman 42(10):L 45–77.

1944b. Recommendations towards conservation of grayfish (dogfish). Fish. Res. Board Can., Pac. Progr. Rep. 61: 24.

1946. The biology of the soupfin shark *Galeorhinus zyopterus* and biochemical studies of the liver. Calif. Bur. Mar. Fish. Fish. Bull. 64: 93 p.

1948. Regulation and investigation of the Pacific halibut fishery in 1947. Rep. Int. Fish. Comm. (Seattle) 13: 35 p.

1949. The commercial fish catch of California for the year 1947 with a historical review 1916–1947. Calif. Dep. Nat. Resources, Div. Fish Game. Fish. Bull. 74: 267 p.

1958. The menace of the dogfish shark on the Pacific coast p. 141–153. *In* 68th Ann. Rep. Wash. Dep. Fish. 303 p.

1964. Catch records of a trawl survey conducted by the International Pacific Halibut Commission between Unimak Pass and Cape Spencer, Alaska from May 1961 to April 1963. Rep. Int. Pac. Halibut Comm. (Seattle) 36: 524 p.

1981. 1981 Fisheries statistical Report. Wash. Dep. Fish., 79 p.

Barraclough, W. E. 1948a. The sunken gill-net fishery and an analysis of the dog-fish (*Squalus acanthias* Girard) and the soup-fin shark (*Galeorhinus galeus* Linnaeus) in British Columbia waters from 1943 to 1946. M.A. thesis, Dep. Zool., Univ. British Columbia, Vancouver, B.C.

1948b. Measures of abundance in dogfish *(Squalus suckleyi)*. Trans. R. Soc. Can. 3(42): 37–43.

BEAGLEHOLE, J. E. [ED.]. 1967. The journals of Captain James Cook. Cambridge Univ. Press.

BEAMISH, R. J., AND M. S. SMITH. 1976. A preliminary report on the distribution, abundance and biology of juvenile spiny dogfish *(Squalus acanthias)* in the Strait of Georgia and their relationship to other fishes. Can. Tech. Rep. Fish. Mar. Serv. 629: 44 p.

BEAMISH, R. J., M. S. SMITH, AND R. SCARSBROOK. 1978. Hake and pollock study, Strait of Georgia cruise G. B. REED, Jan. 6–Feb. 21, 1975. Can. Data Rep. Fish. Mar. Serv. 48: 206 p.

BEAMISH, R. J., M. S. SMITH, V. EGAN, D. BROWN, AND G. MCFARLANE. 1981. Results of spiny dogfish *(Squalus acanthias)* tagging in the Strait of Georgia in 1979. Can. Data Rep. Fish. Aquat. Sci., 262: 73 p.

BEAMISH, R. J., G. A. MCFARLANE, K. R. WEIR, M. S. SMITH, J. R. SCARSBROOK, A. J. CASS, AND C. C. WOOD, 1982. Observations on the biology of Pacific hake, walleye pollock and spiny dogfish in the Strait of Georgia, Juan de Fuca Strait and off the west coast of Vancouver Island and the United States, July 13–24, 1976. Can. MS Rep. Fish. Aquat. Sci. 1651: 150 p.

BIGELOW, H. B., AND W. C. SCHROEDER. 1948. Lancelets, cylostomes and sharks. Sharks, p. 59–576. *In* Fishes of the western North Atlantic. Mem. Sears Found. Mar. Res. 1(1): 1–588.

BONHAM, K. 1954. Food of the dogfish *Squalus acanthias*. Wash. Dep. Fish Res. Papers 1(1): 25–36.

BONHAM, K., F. B. SANFORD, W. CLEGG, AND G. C. BRUCKER. 1949. Biological and Vitamin A studies of dogfish *(Squalus suckleyi)* landed in the State of Washington. Wash. Dep. Fish. Biol. Rep. 49A: 83–114.

BOWEN, L. 1982. Boss Whistle. Oolichan Books. Lantzville (Canada), 280 p.

BRETT, J. R., AND J. M. BLACKBURN. 1978. Metabolic rate and energy expenditure of the spiny dogfish, *Squalus acanthias*. J. Fish. Res. Board Can. 35: 816–821.

BROCKLESBY, H. N. 1927. Determination of vitamin A content in liver oil of the dogfish *Squalus suckleyi*. J. Can. Chem. Met. 11: 238–239.

[ED.]. 1941. The chemistry and technology of marine animal oils with particular reference to those of Canada. Bull. Fish. Res. Board Can. 59: 442 p.

BROWN, D.R., L.G. VANEGAN, M. S. SMITH, AND R. J. BEAMISH. 1979. Results of spiny dogfish *(Squalus acanthias)* tagging in the Strait of Georgia in 1978. Can. Fish. Mar. Serv. Data Rep. 141: 33 p.

BUDKER, P. 1971. The life of sharks. Columbia Univ. Press, New York, NY, 222 p.

CARLSON, C. 1979. The early component at Bear Cove. Can. J. Archaeol. 3: 177–194.

CLARK, D. H. 1958. Dogfish — Puget Sound Villain. Seattle Times, Magazine Sec. Aug. 17, p. 7.

CHATWIN, B. M., AND C. R. FORRESTER. 1953. Feeding habits of dogfish (*Squalus suckleyi* (Girard)). Fish. Res. Board Can. Pac. Progr. Rep. 95: 35–38.

CHILDS, E. A., J. N. GAFFKE, AND D. L. CRAWFORD. 1972. Exposure of dogfish shark feti to mercury. Bull. Environ. Contam. Toxicol. 9: 276–280.

CLEAVER, F. C. 1951. Fisheries statistics of Oregon. Contrib. No. 16: 176 p.

CLEMENS, W. A., AND G. V. WILBY. 1946. Fishes of the Pacific coast of Canada. 1st ed. Bull. Fish. Res. Board Can. 68: 368 p.

1961. Fishes of the Pacific coast of Canada. 2nd ed. Bull. Fish. Res. Board Can. 68: 443 p.

COMPAGNO, L. J. V. 1973. Interrelationships of elasmobranchs. *In* P. H. Greenwood, R. S. Miles, and C. Patterson. [ed.] Interrelationships of fishes. Zool. J. Linn. Soc. 53, Suppl 1.

1984. FAO species catalogue. Vol. 4. Sharks of the world. Part 1. Hexanchiformes and Lamniformes. FAO Fish. Synop., (125) Vol. 4, Pt. 1: 249 p.

COOPER, J. 1874. Remarks on the fisheries of British Columbia, by the agent of the Department of Marine and Fisheries in Victoria. Can. Dep. Mar. and Fish., 6th Ann. Rep., Fish. Branch, App. V: 205–206.

1876. Report of the agent for British Columbia of the Department of Marine and Fisheries for the fiscal year ended June 30, 1875. Can. Dep. Mar. and Fish. 8th Ann. Rep., App. 8: 101–109.

DAWSON, G. M. 1880. Report on the Queen Charlotte Islands. 1878. Rep. of Explor. and Surveys. Geol. Survey Can., 190 p.

DEWHIRST, J. 1980. The Yuquot project. Vol. 1. The indigenous archaeology of Yuquot, a Nootkan outside village. History and Archaeol. 39: 365 p.

DIDONATO, G. 1974. A review of the Puget Sound dogfish fisheries of the 1940s. *In* Puget Sound dogfish *(Squalus acanthias)* studies. Wash. Dep. Fish., Mar. Fish Invest., Suppl. Progr. Rep. 85 p.

DIDONATO, G., AND D. WESTGARD. 1974. Feasibility study for the utilization of Puget Sound dogfish *(Squalus acanthias)* for purposes of incorporation as dogfish meal in the fish diets of the State's salmon and trout hatcheries. Wash. Dep. Fish., Progr. Rep. (October, 1974), 51 p.

FAVORITE, F., A. J. DODIMEAD, AND K. NASU. 1976. Oceanography of the subarctic Pacific region. Bull. Int. North Pac. Fish. Comm. 33: 187 p.

FLADMARK, K. 1971. New radiocarbon dates may push back history in Queen Charlotte Islands. The Midden. Publ. Archaeol. Soc. B.C. 3(5): 11–15.

1982. An introduction to the prehistory of British Columbia. Can. J. Archaeol. 6: 95–156.

FORD, E. 1921. A contribution to our knowledge of the life histories of the dogfishes landed at Plymouth. J. Mar. Biol. Assoc. U.K. 12: 468–505.

FOERSTER, R. E. 1942. Dogfish tagging — preliminary results. Fish. Res. Board Can., Pac. Progr. Rep. 53: 12–13.

FORRESTER, C. R., K. S. KETCHEN, AND C. W. WONG. 1972. Mercury content of spiny dogfish *(Squalus acanthias)* in the Strait of Georgia, British Columbia. J. Fish. Res. Board Can. 29: 1487–1490.

FUJIOKA, B. PATTIE, AND G. DIDONATO. 1974. Dogfish tagging studies in Washington waters. *In* Puget Sound dogfish *(Squalus acanthias)* studies. Wash. Dep. Fish, Mar. Fish. Invest., Suppl. Progr. Rep. 85 p.

GORSHKOV, S. G. [ED.]. 1976. World ocean atlas. Vol. 1. Pacific Ocean. Pergramon Press, Oxford, 302 p.

GULLAND, J. A. 1970. The fish resources of the oceans. F.A.O. Fish. Tech. Pap. 97: 425 p.

1983. Fish stock assessment (Vol. 1). A manual of basic methods. John Wiley and Sons, New York, NY. 223 p.

HALL, A. S., F. M. TEENY, AND J. GAUGLITZ. 1977. Mercury in fish and shellfish of the Northeast Pacific. III. Spiny dogfish *(Squalus acanthias)*. Fish. Bull. (Seattle), 75(3): 642–645.

HANAVAN, M. G., AND G. K. TANONAKA. 1959. Experimental fishing to determine distribution of salmon in the North Pacific Ocean, 1956. U.S. Dep. Inter. Fish Wildl. Serv., Spec. Sci. Rep. 302: 22 p.

HARRISON, R. W., AND V. J. SAMPSON. 1942a. The Pacific coast shark and dogfish liver fisheries. Part 1, Pac. Fisherman, 40(8): 29–31. 1942b. The Pacific coast shark and dogfish liver fisheries. Part 2, Pac. Fisherman 40(9): 37–39.

HART, J. L. 1942. Reproduction in the dogfish. Fish. Res. Board Can., Pac. Progr. Rep. 51: 16–17.

1946. Relationship between value and size of dogfish.

Fish. Res. Board Can. Pac. Progr. Rep. 66: 16.

1973. Pacific fishes of Canada. Bull. Fish. Res. Board Can. 180: 740 p.

HOLDEN, M. J. 1966. The food of the spurdog, *Squalus acanthias* L. J. Cons. Cons. Int. Explor. Mer 30: 255–266.

1977. Elasmobranchs, p. 187–214. *In* J. A. Gulland [ed.] Fish population dynamics. J. Wiley and Sons Ltd., London. 372 p.

HOLDEN, M. J., AND P. S. MEADOWS. 1964. The fecundity of the spurdog (*Squalus acanthias* L.). J. Cons. Perm. Int. Explor. Mer 28: 418–424.

HOLLAND, G. A. 1957. Migration and growth of the dogfish shark, *Squalus acanthias* (Linnaeus) of the eastern North Pacific. Wash. Dep. Fish., Res. Pap. 2(1): 43–59.

HOURSTON, A. S., AND C. W. HAEGELE. 1980. Herring on Canada's Pacific coast. Can. Spec. Publ. Fish. Aquat. Sci. 48: 23 p.

IRELAND, W. E. 1954. The historical evolution of the present settlement pattern. Trans. 7th B.C. Nat. Resour. Conf., p. 197–201.

JONES, B. C., AND G. H. GEEN. 1976. Taxonomic reevaluation of the spiny dogfish (*Squalus acanthias* L.) in the northeastern Pacific Ocean. J. Fish. Res. Board Can. 33: 2500–2506.

1977. Food and feeding of spiny dogfish *(Squalus acanthias)* in British Columbia waters. J. Fish. Res. Board Can. 34: 2067–2078.

KATO, S., S. SPRINGER, AND M. H. WAGNER. 1967. Field guide to eastern Pacific and Hawaiian sharks. U.S. Dep. Inter. Fish. Wildl. Circ. 471: 47 p.

KAUFFMAN, D. E. 1955. Noteworthy recoveries of tagged dogfish. Wash. Dep. Fish., Fish. Res. Pap. 1(3): 39–40.

KENNEDY, W. A. 1972. Preliminary study of sablefish culture, a potential new industry. J. Fish. Res. Board Can. 29: 207–210.

KETCHEN, K. S. 1969. A review of the dogfish problem off the west coast of Canada. Fish. Res. Board Can. MS Rep. 1048: 25 p.

1972. Size at maturity, fecundity and embryonic growth of the spiny dogfish *(Squalus acanthias)* in British Columbia waters. J. Fish. Res. Board Can. 29: 1717–1723.

1975. Age and growth of dogfish *(Squalus acanthias)* in British Columbia waters. J. Fish. Res. Board Can. 32: 43–59.

LANGEVIN, H. L. 1872. Extract from an official report on British Columbia by the Minister of Public Works. Can. Dep. Mar. and Fish., Ann. Rep. for the year ending 30th June, 1871. App. Q. 176–180.

LEAMAN, J. E. 1982. Catch and effort statistics of the Canadian groundfish fishery on the Pacific coast in 1981. Can. Tech. Rep. Fish. Aquat. Sci. 1124: 90 p.

1983. Catch and effort statistics of the Canadian groundfish fishery on the Pacific coast in 1982. Can. Tech. Rep. Fish. Aquat. Sci. 1226: 88 p.

LEIM, A. H., AND W. B. SCOTT. 1966. Fishes of the Atlantic coast of Canada. Bull. Fish. Res. Board Can. 155: 485 p.

LESLIE, P. H., AND D. H. S. DAVIS. An attempt to determine the absolute number of rats on a given area. J. Anim. Ecol. 8: 94–113.

LORD, J. K. 1866. The naturalist in Vancouver Island and British Columbia. Vol. 1, Richard Bentley, London, 358 p.

LYONS, C. 1969. Salmon and our heritage. Mitchell Press Ltd., Vancouver, B.C. 768 p.

MCEACHERN, D. B., AND R. ROBERTS. 1973. An analysis of the British Columbia dogfish subsidy program, February–March, 1973. Environ. Can., Fish Mar. Serv. MS 23 p. (Mimeo.)

MCFARLANE, G. A., R. J. BEAMISH, M. S. SMITH, V. EGAN, AND D. BROWN. 1982. Results of spiny dogfish *(Squalus acanthias)* tagging in the Strait of Georgia, Queen Charlotte Sound, Hecate Strait and Dixon Entrance, during 1980. Can. Fish. Aquat. Sci. MS Rep. 1646: 123 p.

MANZER, J. I. 1946. Interesting movements as shown by the recovery of certain species of tagged fish. Fish. Res Board Can., Pac. Progr. Rep. 67: 31.

MASTROMATTEO, E., AND R. B. SUTHERLAND. 1972. Mercury in humans in the Great Lakes Region, p. 86–92. *In* R. H. Harting and B. D. Dinman [ed.]. Environmental mercury accumulation. Ann Arbor Publ. Inc., Ann Arbor, MI. 349 p.

MOTHERWELL, J. A. 1922. Report of the chief inspector, western fisheries division (British Columbia) for 1921. Can. Dep. Mar. and Fish., 55th Ann. Rep., Fish. Branch, App. 1: 30–55.

1923. Report of the chief inspector, western fisheries division (British Columbia) for 1922. Can. Dep. Mar. and Fish. 56th Ann. Rep. Fish. Branch, App. 1: 25–60.

1946. Annual report of chief supervisor of fisheries, western division (British Columbia) for 1944. Can. Dep. Fish. 15th Ann. Rep., App. 1: 19–48.

MOWAT, T. 1888. Annual report on the fisheries of British Columbia for the year 1887. Can. Dep. Fish. Ann. Rep., App. 7: 239–268.

1890. Annual report on the fisheries of British Columbia for the year 1889. Can. Dep. Fish., Ann. Rep., App. 9: 247–262.

OKADA, S., AND K. KOBAYASHI. 1968. Colored illustrations of pelagic and bottom fishes in the Bering Sea. (Spec. Publ.) North Pac. Sal. Res. Assoc. and Japan Suisan Resource Protection Assoc. 179 p.

OKADA, Y. 1966. Fishes of Japan. Uno Shoten Co. Ltd., Tokyo, 458 p.

PATTIE, B., AND J. TAGART. 1983. The 1981 Washington trawl landings by Pacific Marine Fisheries Commission and bottomfish statistical areas. Wash. Dep. Fish. Progr. Rep. 37: 44 p.

PEDERSEN, M. 1980. Review of the set net fisheries for groundfish in Puget Sound, Washington, 1974–76. Wash. Dep. Fish. Progr. Rep. 113: 33 p.

1981. Review of the set net fisheries of Washington State during 1977–1979. Wash. Dep. Fish. Progr. Rep. 138: 30 p.

PEDERSEN, M., AND G. DIDONATO. 1982. Groundfish management plan for Washington's inside waters. Wash. Dep. Fish. Progr. Rep. 170, 123 p.

PITTENDRIGH, G. 1886. Annual report on the fisheries of British Columbia. Can. Dep. Fish., App. 8: 273–297.

POWELL, D. E., AND A. E. PETERSON. 1957. Experimental fishing to determine distribution of salmon in the North Pacific Ocean, 1955. U.S. Dep. Inter. Fish Wildl. Serv. Spec. Sci. Rep. 205: 30 p.

PRATT, J. H. 1904. Report of the Fisheries District No. 1, New Brunswick, comprising the counties of Charlotte and St. John for the year 1903. App. 4: 96–101. *In* 37th Ann. Rep., Dep. Mar. and Fish. for 1904, 355 p.

PUGSLEY, L. I. 1937. Variations in the vitamin A content of the liver oil of the greyfish *(Squalus suckleyi)*. Fish. Res. Board Can. Pac. Progr. Rep. 31: 3–5.

1940. Factors influencing the vitamin A and D potency of grayfish liver oil, *Squalus suckleyi* (Girard). J. Fish. Res. Board Can. 4(5): 312–322.

ROBINSON, C. K., L. A. LAPI, AND E. W. CARTER. 1982. Stomach contents of spiny dogfish *(Squalus acanthias)* caught near the Qualicum and Fraser rivers, April–May, 1981–82. Can. MS Rep. Fish. Aquat. Sci. 1656: 21 p.

ROEDEL, P. M., AND W. E. RIPLEY. 1950. California sharks and rays. Calif. Dep. Nat. Res., Fish and Game Div., Fish Bull. 75: 88 p.

SANFORD, F. B., AND K. BONHAM. 1946. Grayfish liver colour related to fin-spine length. U.S. Dep. Inter. Fish Wild. Serv., Comm. Fish. Rev. 8(6): 6–8.

SANFORD, F. B., G. A. HOLLAND, AND G. C. BUCHER. 1950. Vitamin A in 155 grayfish livers. U.S. Dep. Inter. Fish Wildl. Serv., Comm. Fish. Rev. 12(3): 17–21.

SHEPARD, M. P., AND J. C. STEVENSON. 1956. Abundance, distribution and commercial exploitation of the fisheries resources of Canada's west coast. Trans. 9th B.C. Nat. Res. Our. Conf., p. 131–190.

SHMIDT, P. YU. 1950. Fishes of the Sea of Okhotsk. 392 p. (Translated from Russian by Israel Program for Sci. Translations, Jerusalem, 1965).

SMITH, J. E. 1972. Catch and effort statistics of the Canadian groundfish fishery on the Pacific coast in 1971. Fish. Res. Board Can. Tech. Rep. 317: 66 p.

1981. Catch and effort statistics of the Canadian groundfish fishery on the Pacific coast in 1980. Can. Tech. Rep. Fish. Aquat. Sci. 1032: 90 p.

SMITH, R. T. 1936. Report on the Puget Sound otter trawl investigations. Wash. Dep. Fish. Biol. Rep. 36B: 61 p.

SPALDING, D. J. 1964. Comparative feeding habits of the fur seal, sea lion and harbour seal on the British Columbia coast. Bull. Fish. Res. Board Can. 146: 52 p.

STEWART, H. 1979. Looking at Indian art of the northwest coast. Douglas and McIntyre, Vancouver, B.C., 111 p.

1984. Cedar. Tree of life to the northwest coast Indians. Douglas and McIntyre, Vancouver, B.C. 192 p.

STOCKER, M. 1981. Groundfish stock assessments off the west coast of Canada in 1981 and recommended total allowable catches for 1982. Can. MS Rep. Fish. Aquat. Sci. 1626: xxx + 282 p.

SUTTLES, W. P. 1951. Economic life of the coast Salish of Haro and Rosario Straits, 472 p. *In* Coast Salish and Washington Indians. Garland Publishing Inc. (1974)., 512 p.

SWAIN, L. A. 1944. The Pacific coast dogfish shark liver oil industry. Fish. Res. Board Can., Pac. Progr. Rep. 58: 3–7.

SWANTON, J. R. 1905. Contributions to the ethnology of the Haida. Jesup North Pac. Exped., Memoir Amer. Mus. Nat. Hist., New York, V(1), 300 p.

SWORD, C. B. 1905. Report on the fisheries of British Columbia for the year 1903. Can. Dep. Mar. and Fish., 37th Ann. Rep., App. 2: 29–47.

TAYLOR, E. G. 1908. Report on the fisheries of British Columbia for the year 1907–08 (District 3). Can. Dep. Mar. and Fish., 41st Ann. Rep., App. 11: 220–221.

1916. Report on the fisheries of District 3 (British Columbia). Can. Dep. Naval Serv. for 1915–16. Fish. Branch, 49th Ann. Rep., App. 9: 243–299.

1917. Report on the fisheries of District 3 (British Columbia). Can. Dep. Naval Serv., Fish. Branch, 50th Ann. Rep., App. 9: 230–273.

TEMPLEMAN, W. 1963. Distribution of sharks in the Canadian Atlantic (with special reference to Newfoundland waters). Bull. Fish. Res. Board Can. 140: 77 p.

THOMSON, J. A., AND A. N. YATES. 1961a. British Columbia landings of trawl-caught groundfish by month, by minor statistical area. Vol 4: 1945–1947. Fish. Res. Board Can. Stat. Circ. Ser. No. 4.

1961b. British Columbia landings of trawl-caught groundfish by month, by minor statistical area. Vol. 5: 1948–1950. Fish. Res. Board. Can., Stat. Circ. Ser. No. 5.

WALSH, S. J. 1984. Atlantic spiny dogfish. Underwater World. Communications Directorate, Can. Dep. Fish. and Oceans, Ottawa, 6 p.

WEBB, P. W. 1975. Hydrodynamics and energetics of fish propulsion. Bull. Fish. Res. Board Can. 190: 159 p.

WEDIN, J. H., AND M. E. MOORE. 1959. The menace of the dogfish shark on the Pacific coast, p. 20–31. *In* U.S. Senate Committee Hearings on Dogfish Shark Eradication, Bill S. 1264. Univ. Wash., Tech. File No. 1930.

WHITCHER, W. F. 1877. Report of the Commissioner of Fisheries. Can. Dep. Mar. and Fish., 9th Ann. Rep., Suppl. 4: 45 p.

WOOD, C. C., K. S. KETCHEN, AND R. J. BEAMISH. 1979. Population dynamics of spiny dogfish *(Squalus acanthias)* in British Columbia waters. J. Fish. Res. Board Can. 36: 647–656.

Appendix 1

Numbers of American tagged spiny dogfish by areas of release and areas of recapture. Tagging period: May 1969 to November 1972; recapture period: (South and Central Sounds) May 1969 to January 1974; (North Sound and Swiftsure (May 1969 to 1983)[a]. See Fig. 3 for area code references.

		Areas of recapture													
		Puget Sound				Strait of Georgia			Juan de Fuca		Outside areas				
		South Sound	Central Sound	Hood Canal	North Sound	Gulf Is.	Fraser R.	North Strait	U.S. side	Can. side	Swiftsure Bank	Other 3C	Other areas	UNK.	Total all areas
Areas of release	No. tagged	85	83,84	88	80	17–19	28–29	13–16	86,89	20	21	23,24			
South Sound (85)	9 155	252	99	1	2	—	—	—	2	—	1	—	—	4	361
Central Sound (83,84)	5 160	36	164	3	8	—	—	—	16	—	2	—	—	2	231
North Sound (80)	3 013	—	5	—	147	7	10	6	4	2	1	—	15	4[b]	186
Swiftsure (21)	6 751	2	19	2	8	8	1	2	9	9	12	16	11[c]	9[d]	108
Total	24 079	290	287	6	165	15	11	8	31	11	16	16	26	19	886

[a]Table adapted from Fujioka et al. (1974; table 2) with assistance of M. Pedersen (personal communication) in extending coverage of recapture period to present for taggings noted in heading.

[b]Includes 1 recapture from unknown area within Canadian Strait of Georgia (Area 4B).

[c]Includes 1 — Hecate Strait (Area 5C–5D); 1 — Queen Charlotte Sound (Area 5A–5B); 4 — outer Washington coast (Area 3A–3B); 4 — Oregon (Area 2B–2D), and 1 — California (Area 1C).

[d]Includes 5 recaptures from unknown areas within U.S. inside waters (Area 4A)

Appendix 2

Numbers of Canadian tagged spiny dogfish by areas of release within the Strait of Georgia (1978–80) and by areas of recapture to December 31, 1981[a]. See Fig. 3 for area code references.

			Areas of recapture																		
			Strait of Georgia											Puget Sound				Juan de Fuca		Other	
Area of tagging	Year of tagging	Year of recap.	12	13	14	15	16	17A	17B	18	19	28	29	North Sound	Central Sound	Hood Canal	South Sound	U.S. side	Can. side	Off shore	Total
Area 13	1978	1978	—	3	2	5	2	—	2	—	—	—	1	—	—	—	—	—	—	—	15
		1979	—	3	1	2	1	—	2	—	—	3	1	2	—	—	—	—	—	—	15
		1980	—	2	—	4	—	—	—	—	—	—	—	2	—	—	—	—	—	—	8
		1981	—	—	—	1	—	1	1	—	—	—	—	—	—	—	—	—	—	—	3
		Total	—	8	3	12	3	1	5	—	—	3	2	4	—	—	—	—	—	—	41
	1979	1979	1	12	2	3	1	—	4	—	2	3	—	—	—	—	—	1	—	—	29
		1980	1	17	7	1	1	2	1	1	1	—	—	2	—	—	—	—	—	—	34
		1981	—	17	1	2	2	—	—	—	—	1	—	—	—	—	—	—	—	—	23
		Total	2	46	10	6	4	2	5	1	3	4	—	2	—	—	—	1	—	—	86
	1980	1980	—	4	—	3	—	—	1	—	—	—	1	—	—	—	—	—	—	—	9
		1981	—	8	1	2	—	—	1	1	—	2	1	1	—	—	—	1	—	—	18
		Total	—	12	1	5	—	—	2	1	—	2	2	1	—	—	—	1	—	—	27
Area 16	1979	1979	—	—	—	—	—	—	—	—	—	—	—	—	—	—	—	—	—	—	—
		1980	—	—	2	—	4	—	2	1	—	3	1	—	—	—	—	2	—	1	16
		1981	—	2	—	—	2	—	—	—	—	—	—	1	—	—	—	—	—	—	5
		Total	—	2	2	—	6	—	2	1	—	3	1	1	—	—	—	2	—	1	21
Area 17A	1979	1980	—	—	1	—	1	—	—	—	—	—	—	2	—	—	—	—	—	—	4
		1981	—	1	—	—	1	—	—	1	—	—	—	—	—	—	—	—	—	—	3
	1980	1980	—	1	1	—	1	2	—	3	—	1	4	17	3	1	1	1	—	1	37
		1981	—	—	—	—	—	3	1	1	—	5	2	3	—	—	—	1	—	2	18
		Total	—	2	2	—	3	5	1	5	—	6	6	22	3	1	1	2	—	3	62
Area 17B	1980	1980	—	—	—	—	—	—	—	—	—	—	—	1	—	—	—	—	—	—	1
		1981	—	—	—	—	—	—	—	—	—	—	—	1	—	—	—	—	—	—	1
		Total	—	—	—	—	—	—	—	—	—	—	—	2	—	—	—	—	—	—	2
Area 18	1980	1980	—	1	—	—	—	—	2	3	—	—	—	2	—	—	—	2	1	—	11
		1981	—	—	2	—	—	—	—	3	1	—	—	7	—	—	—	1	—	—	14
		Total	—	1	2	—	—	—	2	6		—	—	9	—	—	—	3	1	—	25
Grand Total	1979–80	1979–81	2	71	20	23	16	8	17	14	4	18	11	41	3	1	1	9	1	4	264

[a]Source: Brown et al. (1979); Beamish et al. (1981); McFarlane et al. (1982); G. A. McFarlane (unpublished data).

Appendix 3

Calculation of Equivalent Round Weight of Spiny Dogfish Caught and Landed

During the long history of the Canadian commercial fishery for spiny dogfish, reported landings have appeared in a variety of forms that require conversion to a common measure in assembling a long-term picture of production. From 1876 to 1916 published figures were usually given in terms of only processed product, namely, gallons of liver oil, liver and body oil, or unspecified oil. From 1917 to 1937 weights of whole (round) fish were the standard measure, but as liver increased in market value, more and more landings consisted of weights of liver only. By 1945 all production was recorded in this manner and continued that way through 1965. In more recent years the only measure of the total amount of dogfish processed in Canada was that derived from reported export weight of products: belly flaps, backs, fins and tails. Thus it has been necessary to develop a number of correction factors to express all figures in terms of equivalent round weight caught and landed.

1. Liver Oil to Round (Whole) Weight

From information provided by Brocklesby (1941), the specific gravity of dogfish liver oil at room temperature is estimated as 0.9100. Since 1 L of water weighs 1 kg, it follows that a litre of liver oil weighs 0.9100 kg.

Under laboratory conditions, the oil in liver of dogfish of about 85 cm average length represents about 72% of the liver weight (Brocklesby 1941). However, under conditions of commercial production during years when reasonably complete records were available (1944–50), it appears that the proportion was somewhat lower, averaging 67%. This latter figure has been selected in estimating the weight of liver from weight of liver oil.

Various authors including current and past personnel of the Economics Branch of the Canadian Fisheries Service have used a variety of estimates of the ratio of liver weight to round weight of dogfish ranging from 10 to 15%. The extremities reflect more a convenience of calculation than actual observed values. In fact, close examination of data provided by Bonham et al. (1949) shows that although there is considerable variation from fish to fish depending on sex, size, and maturity stage a reasonable working average is in the vicinity of 11%. We have selected 11.3% for the present study. Thus the conversion factor from liver to round (whole) weight as used here is 8.85.

In summary then, given data on the number of litres of liver oil (X) reported, we may estimate the equivalent round weight (Y) in tonnes as:

$$(1) \qquad Y = \frac{0.9100 \times 8.85 \times X}{1000 \times 0.67}$$

2. Mixed Liver Oil and Body Oil to Round (Whole) Weight

Using published figures on the weight of dogfish landed and the amount of mixed liver oil and body oil recovered and marketed during the period 1930–37, we arrive at a recovery figure of 13.0% by weight for the combined oils. Thus, the above equation is modified as:

$$(2) \qquad Y = \frac{0.9100 \times X}{1000 \times 0.13}$$

3. Calculation of Minimum and Maximum Equivalent Round Weights

For nearly 70 years, i.e. from the beginning of record keeping in 1876 to 1939 it was never entirely certain whether the oil data identified with dogfish referred to liver oil or mixed liver and body oil. Worse still, between 1888 and 1939 dogfish oil (of one kind or the other) was included in a general fish oil category. Although uncertainty was relieved to some extent by knowledge that this category excluded data on by-products from salmon canning and oils derived from whales, Pacific sardine and herring reduction operations and offal from other fisheries such as that on halibut, there remained sufficient doubt to require that minimum and maximum limits be placed on the likely production of dogfish oil alone. The maximum estimate for each year was derived from the first of the equations presented above, on the assumption that all unidentifiable oil was dogfish liver oil. The minimum estimate was derived from the second equation and assumed that (1) for the period 1876–87 all dogfish oil, unspecified as to type, was derived from livers and bodies, and (2) for the period 1888 to 1916 all *fish* oil was derived from the same source. From 1917 to 1939, the minimum was obtained directly from the reported landed weight of dogfish. From 1940 onwards there was no ambiguity to the records and all conversions could be made by applying the first equation.

During 1888–1916, the calculated minimum is not necessarily the true minimum because, since dogfish were unidentifiable in any form, actual production theoretically could have been zero for all of the years involved. This is highly improbable, but in any case the period was one of relatively low *maximum* values, so the results are not grossly misleading.

The process of calculating equivalent round weights of dogfish is best explained by giving an example:

In Appendix 4, the reported total production of dogfish oil was 825 000 L in 1876. Applying equation 1 (assuming liver oil only) we get a round weight estimate of

9917 t. This appears as the maximum estimate in Table 3, and equation 2 gives us the minimum estimate of 5775 t. Data for 1876 through 1916 in Appendix 4 form the basis for calculations that are provided in Table 3.

Similarly Appendix 5 covering the period 1917–39 is the source of all calculated round weights given in Table 4 but with some important differences. In the first place certain data in the former have been deleted from the latter. Fish oil reported from District 1 (Vancouver and Fraser River ports) during 1917–29 is known to have been mostly salmon oil derived from cannery offal. Data for District 2 (northern B.C. ports) in Appendix 4 show large quantities of oil for 1922–29 that cannot be accounted for. They exclude weights of Pacific sardine, herring, eulachon, and whale oils (which were reported separately during most of the period) and bear no relation to fluctuations in the amount of salmon landed for canning. It is known that the north coast, including the Queen Charlotte Island fishery for dogfish was dormant except as noted for 1928 and 1929. The fact that alleged landings ceased abruptly from 1930 onwards, with change in format of reporting, casts doubt on the authenticity of the fish oil records for 1922–29, so accordingly they have been dropped.

Second, from 1930 to 1939 an active fishery for dogfish was confined almost entirely to the Strait of Georgia (Appendix 5). Landings of dogfish were in round weight and therefore required no conversion, except to the metric system to obtain an estimate of the *minimum* weight landed. To obtain a *maximum* albeit improbable estimate, the amounts of fish oil marketed from 1922 to 1929 and the amounts of dogfish oil marketed from 1930 to 1939 were assumed to be liver oil only, and converted to equivalent round weights using equation 1. In 4 years when this treatment yielded estimates that were less than the minimum estimates it was presumed there was but a single estimate for those years (see Table 4).

From 1940 to 1944 in District 3 dogfish were landed in the round, as livered carcasses and as livers only (Appendix 7). However, appropriate corrections have been made to avoid any duplications in calculating total round weights.

4. Liver Weight to Round (Whole) Weight

From 1940 in Districts 1 and 2 and from 1945 in District 3 until at least 1963 the liver was the only part of the dogfish that was landed. To calculate the equivalent round weight Canadian and United States statistics were converted to metric figures and multiplied by the 8.85 factor mentioned above. Data presented in Tables 12, 13, 15, 17, and 19 were derived in this manner.

5. Other Body Parts to Round (Whole) Weight

With the development of the foodfish fishery from 1964 onwards (but particularly after 1976) there were instances where the total round weight of dogfish landed in Canada was not available. However, figures on the weight *processed* in Canada could be derived indirectly from reports on the total products exported, knowing that dogfish backs, bellyflaps and fin-tails were 30, 6.5, and 2% of the round weight, respectively. Whichever of these factors produced the highest estimate of round weight was used in computing some of the Canadian figures from 1977 onwards in Tables 12, 13, 19, and 22.

Figures on the amount of dogfish landed in Canada but trans-shipped to processing plants in Washington State or delivered directly by Canadian vessels to those plants were obtained from *Fishery Market News* a thrice-weekly publication of the National Marine Fisheries Service (Seattle). This contains records of dogfish imports from British Columbia collected by the (U.S.) Food and Drug Administration.

Appendix 4

Marketed quantities of spiny dogfish liver oil and/or oil from whole dogfish, seals, and porpoises, and/or mixed fish oil presumed to be primarily from dogfish, compared with total fish oil produced as a by-product of other fisheries in British Columbia during 1876–1915. (Figures in thousands of litres.)

	District 2		District 3[a]				B.C. total	
			Dogfish and other oil[b]					
Year	Q.C.I. dogfish liver oil	Q.C.I. fish oil	Exported	domestic	Mixed fish oil[c]	Total fish oil	Dogfish oil	All fish oil
1897	—	—	275	550	—	825	825	825
77	—	—	295	182	—	477	477	477
78	—	—	459	227	—	686	686	686
79	59	—	177	295	—	472	531	531
1880	91	—	200	341	—	541	632	62
81	123	—	236	409	—	645	768	768
82	168	—	459	432	—	891	1059	1059
83	182	—	223	764	—	987	1169	1169
84	205	—	—	45	—	45	250	250
1885	182	—	—	95	—	95	277	277
86	91	—	—	114	—	114	205	205
87	182	—	—	305	—	305	487	487
88	—	91	—	—	73	73	—	291
89	—	136	—	—	9	9	—	641
1890	—	168	—	—	68	68	—	736
91	—	305	—	—	164	164	—	1136
92	—	318	—	—	205	205	—	1164
93	—	332	—	—	173	173	—	782
94	—	305	—	—	145	145	—	650
1895	—	282	—	—	145	145	—	614
96	—	141	—	—	68	68	—	282
97	—	159	—	—	73	73	—	436
98	—	182	—	—	105	105	—	568
99	—	186	—	—	100	100	—	659
1900	—	191	—	—	100	100	—	582
01	—	132	—	—	145	145	—	691
02	—	159	—	—	177	177	—	736
03	—	173	—	—	286	286	—	1018
04	—	77	—	—	314	314	—	877
1905	—	59	—	—	336	336	—	800
06	—	73	—	—	332	332	—	568
07	—	36	—	—	341	341	—	532
08	—	109	—	—	332	332	—	614
09	—	136	—	—	396	396	—	955
1910	—	159	—	—	59	59	—	350
11	—	136	—	—	45	45	—	346
12	—	159	—	—	45	45	—	391
13	—	132	—	—	*[d]	*	—	655
14	—	0	—	—	*	*	—	186
1915	—	32	—	—	*	*	—	155
16	—	73	—	—	36	36	—	155

[a]That portion of the district along the east coast of Vancouver Island from Queen Charlotte Strait south to Victoria and the mainland shore of the Strait of Georgia south to Gower Point (i.e. excluding District 1 — Howe Sound, Vancouver, the Fraser River Estuary — and that portion of District 3 encompassing the west coast of Vancouver Island.

[b]From 1876 to 1887 dogfish was the primary source of oil, but was reported as mixed with amounts of seal and porpoise oil.

[c]From 1888 to 1916, it is presumed that in the particular regions of Districts 2 and 3 considered, dogfish remained the principal contributor. Oil from other sources, e.g. salmon cannery offal was negligible as there were few if any canneries in the Strait of Georgia part of District 3 until well into the 1900s, and herring oil had not yet become a marketable product.

[d]In inspectors' reports for 1913–15 a fish-oil category was missing from District 3 tabulations. However dogfish oil production appeared to be at a low level because of weak markets and complaints about the dogfish nuisance were noted at that time in annual reports.

Appendix 5

Spiny dogfish caught and landed by district showing (a) for 1917–29, amounts of fish oil and fish meal marketed and (b) for 1930–39, amounts of dogfish oil and dogfish meal marketed. (Landed dogfish in tonnes; oil in thousands of liters; meal in tonnes.)

							District 3								
	District 1			District 2			Strait of Georgia			West coast			All districts		
	Dogfish	Fish marketed		Dogfish	Fish marketed		Dogfish	Fish marketed		Dogfish	Fish marketed		Dogfish	Fish marketed	
Year	landed	oil	meal	landed	oil	meal	landed	oil	meal	landed	oil	meal	landed	oil	meal
1917	—	136	328	—	—	—	508	67	512	—	—	—	508	203	840
18	—	41	159	—	—	—	2665	202	1476	—	—	—	2665	243	1635
19	—	54	167	—	—	—	2310	192	1509	—	3	—	2310	249	3311
1920	—	64	233	—	—	—	680	99	170	—	90	109	680	253	512
21	—	36	—	—	—	—	2384	167	326	—	—	—	2384	203	326
22	—	56	198		149	378	1825	37	196	—	1	—	1825	343	772
23	—	36	163		403	109	2206	309	474	—	71	—	2206	819	746
24	—	33	104		572	807	3357	494	639	—	—	—	3357	1099	1550
1925	—	36	186		439	585	3358	702	894	—	—	—	3358	1177	1665
26	—	32	193		497	743	3555	458	654	—	62	280	3555	987	1870
27	—	—	—		1008	1353	5112	661	1034	—	36	19	5112	1705	2406
28	—	—	—	6	709	1621	10452	1157	1674	—	4	24	10452	1866	3319
29	—	114	508	1506	656	946	11773	1320	1835	—	—	—	13279	2090	3289
1930	—	—	—	—	—	—	4476	521	816	—	—	—	4476	521	816
31	—	—	—	—	—	—	5096	774	916	—	—	—	5096	774	916
32	—	—	—	—	—	—	1271	160	239	—	—	—	1271	160	239
33	—	—	—	—	—	—	3661	534	713	—	—	—	3611	534	713
34	—	—	—	—	—	—	5308	927	1029	—	—	—	5308	927	1029
1935	—	—	—	—	—	—	3484	556	697	—	—	—	3484	556	697
36	—	—	—	—	—	—	5268	748	966	—	—	—	5268	748	966
37	—	—	—	—	34	—	5136	566	756	—	—	—	5136	600	756
38	—	33	—	—	33	—	7243	898	1407	696	64	—	7243	964	1407
39	—	8	—	710	54	71	4498	600	851	—	—	—	5208	662	922

Appendix 6

The Problem of Assigning Landings of Spiny Dogfish to Area of Capture

Until about 1938 fishing for spiny dogfish occurred within convenient running distance of the several reduction plants operating in the Strait of Georgia. Total catches made in the Strait were more or less equivalent to the landings reported for District 3. However with the opening of the market for livers, vessels (first halibut longliners followed by trawlers and gillnetters) could deliver from farther afield. Vessels fishing in Hecate Strait and Queen Charlotte Sound (District 2), the open coast and even the Strait of Georgia portion of District 3 increasingly made landings at Vancouver. As shown in the accompanying table, after 1937 the percentage of landings assignable to fishing area declined from 100 to 27% by 1944. Special study of the sunken gillnet (set-net) fishery in Hecate Strait and off the west coast of Vancouver Island, beginning in 1943; introduction of a program to obtain logbook records of trawler operations beginning in 1945, and placement of an FRB contact man at the port of Vancouver in 1946, improved coverage of the trawl and longline fisheries during 1945–47, but with the collapse of the gillnet fishery in the post-war years the proportion declined again to a low level. Many of the smaller vessels (like those of the gillnet fleet) delivered to buyers at floating facilities located near fishing grounds and serviced by packers that landed large mixed loads of liver at Vancouver. Introduction of a sales slip system by the Canadian Department of Fisheries, to document transactions between fishermen and purchasers, became operational in 1953 and for most years thereafter coverage became much more effective, though never completely so because submission of sales slips was not mandatory.

In the late 1970's a new problem developed as a result of U.S. demand for round or dressed dogfish. Substantial amounts of dogfish landed at Vancouver (and not documented by sales slips) were transhipped by truck to processing plants at Washington ports. Coverage of these transactions is explained in Appendix 3 (Section 5).

Proportion of total Canadian landings of spiny dogfish (equivalent round weight in tonnes) that could be assigned to area of capture.

Year	Strait of Georgia	West Coast Vancouver Island	Queen Charlotte Sound	Hecate Strait	Total landing known areas	Total landing	% allocated
1935	3484	—	—	—	3484	3484	100.0
36	5268	—	—	—	5268	5268	100.0
37	5135	—	—	410	5545	5545	100.0
38	7243	—	—	22	7265	7640	95.1
39	4498	—	—	538	5036	5139	98.0
1940	5932	—	—	801	6733	8790	76.6
41	6138	—	—	1333	7471	13965	53.5
42	4554	—	—	3142	7696	17027	45.2
43	3582	—	—	2136	5718	20567	27.8
44	1102	556	—	6731	8389	31187	26.9
1945	523	1225	—	10312	12060	23373	51.6
46	658	1832	557	4466	7513	11417	65.8
47	747	699	1317	3630	6393	15089	42.4
48	399	620	986	1528	3533	12178	29.0
49	510	615	681	1090	2896	16010	18.1
1950	314	178	121	57	671	2213	30.3
51	484	286	481	75	1326	4000	33.2
52	638	302	258	18	1216	3053	39.8
53	658	435	223	38	1354	3115	43.5

Appendix 7

Landed quantities of spiny dogfish and liver, and marketed quantities of products, by districts of British Columbia (weights in tonnes, except for oil in thousand of litres).

	District 1		District 2					District 3					All districts				
		Marketed			Marketed					Marketed					Marketed		
Year	Dogfish liver landed	liver[a] oil	Dogfish landed	Dogfish liver landed	Liver oil	Body oil	Meal	Dogfish landed	Dogfish liver landed	Liver oil	Body oil	Meal	Dogfish landed	Dogfish liver landed	Liver oil	Body oil	Meal
1940	286	100	—	18	9	—	—	6100	406	264	432	981	6100	710	373	432	981
41	578	340	338	150	118	14	38	6148	850	559	314	915	6486	1578	1017	328	953
42[b]	1305	955	—	355	255	—	—	4560	264	218	191	829	4560	1924	1428	191	829
43	1831	1405	—	242	177	—	—	3580	251	191	168	642	3580	2324	1773	168	642
44	2053	1505	—	366	214	—	—	1104	1105	796	82	299	1104	3524	2515	82	299
1945	1366	1041	—	670	491	—	—	—	605	427	—	—	—	2641	1959	—	—
46	753	555	—	425	314	—	—	—	112	68	—	—	—	1290	937	—	—
47	—	—	—	—	—	—	—	—	—	—	—	—	—	1705	1232	—	—
48	—	—	—	—	—	—	—	—	—	—	—	—	—	1376	1059	—	—
49	—	—	—	—	—	—	—	—	—	—	—	—	—	1809	1118	—	—
1950	—	—	—	—	—	—	—	—	—	—	—	—	—	250	173	—	—
51	—	—	—	—	—	—	—	—	—	—	—	—	—	452	314	—	—
52	—	—	—	—	—	—	—	—	—	—	—	—	—	345	250	—	—
53	—	—	—	—	—	—	—	—	—	—	—	—	—	352	—[c]	—	—
54	—	—	—	—	—	—	—	—	—	—	—	—	—	284	—	—	—

[a]Includes oil equivalent of marketed livers, in all 3 Districts.
[b]Beginning in 1942 oils were reported by weight rather than volume. Volume measure has been retained here.
[c]From 1953 onwards the amount was expressed in Vitamin A units rather than by weight.

Appendix 8

Spiny dogfish removals from the Strait of Georgia during the first subsidized fishery, Jan. - Mar. 1959. (Figures in tonnes).

	Minor statistical areas										
	13	14	15	16	17	18	19	28	29	UNK	all areas
January											
Liver wt.	—	4.83	—	0.56	13.66	41.77	1.86	0.05	0.24	52.22	115.19
Whole fish equiv.	—	42.74	—	4.96	120.89	369.66	16.46	0.44	2.12	462.15	1019.42
"Killerboat" wt.	—	63.64	—	0.03	11.61	43.10	—	—	0.29	—	118.67
Total round wt.	—	106.38	—	4.99	132.50	412.76	16.46	0.44	2.41	462.15	1138.09
February											
Liver wt.	0.16	2.79	0.74	—	4.06	12.97	0.35	—	—	17.43	38.50
Whole fish equiv.	1.42	24.69	6.55	—	35.93	114.78	3.10	—	—	154.26	340.73
"Killerboat" wt.	—	128.85	—	1.36	4.73	90.40	—	—	—	—	225.34
Total round wt.	1.42	153.54	6.55	1.36	40.66	205.18	3.10	—	—	154.26	566.07
March											
Liver wt.	0.02	0.94	—	—	0.05	0.30	—	—	0.46	1.49	3.26
Whole fish equiv.	0.17	8.32	—	—	0.44	2.66	—	—	4.07	13.19	28.85
"Killerboat" wt.	0.09	314.82	—	0.05	0.45	1.62	—	—	—	—	317.03
Total round wt.	0.26	323.14	—	0.05	0.89	4.28	—	—	4.07	13.19	345.88
Grand total round wt.	1.68	583.06	6.55	6.40	174.05	622.2	19.56	0.44	6.48	629.60	2050.04

Appendix 9

Canadian landings of spiny dogfish liver during the period 1954–70 (weight in tonnes).[a]

Year	Area 4B Strait of Georgia	Areas 3C + 3D West coast Vancouver Island	Areas 5A + 5B Queen Charlotte Sound	Areas 5C + 5D Hecate Strait	Total all Canadian areas
1954	209	41	1	34	285
55	193	35	9	59	296
56	47	7	3	70	127
57	126	63	9	82	280
58	112	21	7	42	182
59	564	72	10	77	723
1960	388	8	9	88	493
61	252	347	7	63	669
62	46	—	—	—	46
63	19	6	—	—	25
64	111	—	—	—	111
1965	28	—	1	tr	29
1966	61	tr	1	—	61
67	66	—	—	—	66
68	38	—	—	—	38
69	tr	—	—	—	tr
1970	—	—	—	—	—

[a]Source: Unpublished file data, Canadian Dep. Fisheries, Economics Branch.

Appendix 10

Canadian production of spiny dogfish by gear and minor statistical areas (MSA) of the Strait of Georgia and Strait of Juan de Fuca (figures in tonnes)[a], 1972–82.

Minor statistical area and gear	1972	1973	1974	1975	1976	1977	1978	1979	1980	1981	1982
MSA 12 & 13											
Trawl	—	69.03	—	5.71	—	8.30	6.44	33.21	31.21	92.95	93.68
Set-line	—	644.49	52.32	62.13	—	—	—	58.51	88.53	2.78	63.78
Other	—	—	—	0.33	—	—	—	—	—	—	—
Total	—	713.52	52.32	68.17	—	8.30	6.44	91.72	119.74	95.73	157.46
MSA 14											
Trawl	16.38	315.42	20.38	14.29	1.14	67.07	140.48	84.14	86.98	27.58	90.28
Set-line	—	65.76	—		—	45.38	16.00	208.58	177.82	82.33	59.45
Other	—	134.83	—	47.29	20.73	—	—	—	—	—	—
Total	16.38	516.01	20.38	61.58	21.87	112.45	156.48	292.72	264.80	109.91	149.73
MSA 15											
Trawl	—	—	—	—	—	—	—	—	—	—	—
Set-line	—	217.32	—	—	—	195.59	205.00	245.60	116.09	11.43	36.49
Other	—	0.24	—	—	—	9.99	—	—	—	—	—
Total	—	217.56	—	—	—	206.58	205.00	245.60	116.09	11.43	36.49
MSA 16											
Trawl	—	—	—	—	—	—	—	—	36.13	—	—
Set-line	—	696.26	5.14	18.00	63.53	306.77	962.20	410.83	240.48	165.50	283.36
Other	—	18.18	—	21.18	34.13	6.51	—	—	—	—	—
Total	—	714.44	5.14	39.18	97.66	312.28	962.20	410.83	276.61	165.50	283.46
MSA 17											
Trawl	3.48	543.39	89.61	183.09	11.13	59.25	21.64	151.11	44.24	4.98	27.67
Set-line	32.48	1097.31	632.25	21.16	2.73	278.64	560.20	1585.32	285.89	101.23	369.40
Other	—	3.97	—	28.82	11.48	3.48	—	—	1.61	—	—
Total	35.96	1644.67	721.86	233.07	25.34	341.41	581.84	1736.43	331.74	106.21*	397.07
MSA 18											
Trawl	52.87	309.58	183.33	233.58	46.30	377.70	183.88	285.58	149.08	38.17	210.29
Set-line	—	164.16	37.44	12.61	22.24	42.20	207.00	311.25	225.78	27.38	96.84
Other	—	1.08	—	21.17	2.56	—	—	—	—	—	—
Total	52.87	474.82	220.77	267.35	71.10	419.90	390.88	596.83	374.86	65.55	307.13
MSA 19											
Trawl	—	—	9.80	10.36	23.36	135.94	347.36	301.83	194.53	281.28	236.58
Set-line	—	—	0.85	—	—	3.64	51.00	134.56	54.81	13.28	15.22
Other	—	—	—	—	—	—	—	—	—	—	—
Total	—	—	10.65	10.36	23.36	139.58	398.36	436.39	249.34	294.56	251.80
MSA 20											
Trawl	—	—	2.92	—	—	—	19.88	22.92	20.41	0.50	37.35
Set-line	—	—	—	—	—	—	15.00	38.72	101.71	—	—
Other	—	—	—	—	—	—	3.20	—	1.61	—	—
Total	—	—	2.92	—	—	—	38.08	61.64	123.73	0.50	37.35
MSA 28-29											
Trawl	1.09	13.90	0.87	—	—	3.05	0.28	5.39	1.99	1.16	—
Set-line	—	—	—	—	—	51.28	91.00	456.59	245.51	361.76	390.14
Other	—	—	—	0.45	—	—	—	—	—	—	—
Total	1.09	13.90	0.87	0.45	—	—	91.28	461.98	247.50	362.92	390.14
ALL AREAS											
Trawl	73.82	1251.91	306.91	447.03	81.93	651.35	721.96	884.18	564.47	446.62*	695.85
Set-line	32.48	2885.30	728.00	113.90	88.50	923.50	2107.40	3449.96	1536.62	765.69	1314.68
Other	—	158.30	—	119.23	68.90	62.10	3.20	—	3.22	—	—
Total	106.30	4294.92	1034.92	680.16	239.16	1636.95	2830.56	4334.14	2104.41	1212.31	2010.53

[a]Source: Smith (1973) for the year 1972 and similar publications for years 1973–80; Leaman (1982 and 1983) for 1981 and 1982, except for adjustments to MSA 17, 18, 19, 20, 28, and 29 to allow for unrecorded landings of 449 and 755 t, respectively, in the State of Washington.

Appendix 11

Washington State production of spiny dogfish by gear and subareas of Puget Sound and American waters of the Strait of Georgia and Strait of Juan de Fuca (Figures in tonnes)[a].

Subarea and gear	1976	1977	1978	1979	1980
North Sound					
Trawl	446.1	449.9	449.9	833.7	541.3
Set-net	246.5	434.7	310.7	230.5	199.0
Set-line	29.7	151.0	393.8	637.0	931.4
Other	—	—	—	—	—
Total	722.3	1035.6	1154.4	1701.2	1662.7
Straits-Admiralty					
Trawl	8.8	25.4	84.3	105.2	98.9
Set-net	450.1	509.2	475.6	315.7	216.5
Set-line	56.6	64.5	261.0	765.2	356.2
Other	—	—	—	—	—
Total	515.5	599.1	820.9	1186.1	671.6
Hood Canal					
Trawl	—	—	—	—	—
Set-net	871.9	118.3	42.0	194.1	171.0
Set-line	37.0	0.8	68.2	80.2	48.9
Other	—	—	—	—	—
Total	908.9	119.1	110.2	274.3	219.9
Central Sound					
Trawl	76.9	37.8	102.8	85.4	85.5
Set-net	270.4	361.3	155.7	238.0	185.2
Set-line	74.8	111.9	47.9	201.7	111.0
Other	5.4	5.4	4.5	9.3	5.1
Total	427.5	511.4	310.9	534.4	386.8
South Sound					
Trawl	0.4	15.6	40.8	28.9	11.1
Set-net	7.9	81.0	201.3	141.0	34.6
Set-line	—	4.3	8.6	16.9	8.3
Other	—	—	—	—	—
Total	8.3	100.9	250.7	186.8	54.0
All Subareas					
Trawl	532.2	523.7	677.8	1053.2	736.8
Set-net	1846.8	1504.5	1185.3	1119.3	806.3
Set-line	198.1	332.5	779.5	1701.0	1455.8
Other	5.4	5.4	4.5	9.3	5.1
Total	2582.5	2366.1	2647.1	3882.9	3004.0

[a]Source: Pedersen and DiDonato (1982; tables 8, 15, 22, 27, and 33).